Modern Birkhäuser Classics

Many of the original research and survey monographs in pure and applied mathematics published by Birkhäuser in recent decades have been groundbreaking and have come to be regarded as foundational to the subject. Through the MBC Series, a select number of these modern classics, entirely uncorrected, are being re-released in paperback (and as eBooks) to ensure that these treasures remain accessible to new generations of students, scholars, and researchers.

T0178034

For other titles published in this series, go to
www.springer.com/series/7588

George Pólya
Robert E. Tarjan
Donald R. Woods

Notes on Introductory Combinatorics

Reprint of the 1983 Edition

Birkhäuser
Boston · Basel · Berlin

George Pólya (Deceased)
Department of Mathematics
Stanford University
Stanford, CA 94305
USA

Robert E. Tarjan
Department of Computer Science
Princeton University
Princeton, NJ 08544
USA

Donald R. Woods
Google Inc.
1600 Amphitheatre Parkway
Mountain View, CA 94043
USA

Originally published in the series *Progress in Computer Science and Applied Logic*

ISBN 978-0-8176-4952-4   e-ISBN 978-0-8176-4953-1
DOI 10.1007/978-0-8176-4953-1

Library of Congress Control Number: 2009938715

Mathematics Subject Classification (2000): 05A15, 05C45, 05D10, 20Bxx, 68R05

Printed on acid-free paper

Birkhäuser Boston is a part of Springer Science+Business Media (www.birkhauser.com)

George Pólya
Robert E. Tarjan
Donald R. Woods
Notes on Introductory Combinatorics

1983

Birkäuser
Boston · Basel · Berlin

Authors:

George Pólya
Department of Mathematics
Stanford University
Stanford, California 94305, USA

Robert E. Tarjan
Bell Laboratories
600 Mountain Avenue
Murray Hill, New Jersey 07974, USA

Donald R. Woods
Xerox Corporation
3333 Coyote Hill Road
Palo Alto, California 94304, USA

Library of Congress Cataloging in Publication Data

Pólya, George, 1887-
      Notes on introductory combinatorics.

      (Progress in computer science ; no. 4)
      Bibliography: p.
      1. Combinatorial analysis.   I. Tarjan, Robert E.
(Robert Endre), 1948-   . II. Woods, Donald R.,
1954-   . III. Title.   IV. Series.
QA164.P635   1983   511'.6   83-15790
ISBN 0-8176-3123-2
ISBN 0-8176-3170-4 (pbk.)

CIP-Kurztitelaufnahme der Deutschen Bibliothek

Pólya, George:
Notes on introductory combinatorics / George
Pólya ; Robert E. Tarjan ; Donald R. Woods. -
Boston ; Basel ; Stuttgart : Birkhäuser, 1983.

      (Progress in computer science ; No. 4)
      ISBN 3-7643-3170-4 (Basel, Stuttgart) brosch.
      ISBN 0-8176-3170-4 (Boston) brosch.
      ISBN 3-7643-3123-2 (Basel, Stuttgart) Pp.
      ISBN 0-8176-3123-2 (Boston) Pp.

NE: Tarjan, Robert E.:; Woods, Donald R.:; GT

© Birkhäuser Boston, Inc., 1983
ISBN 0-8176-3123-2 (hardcover); 0-8176-3170-4 (paperback)
ISBN 3-7643-3123-2 (hardcover); 3-7643-3170-4 (paperback)
Printed in USA

9 8 7 6 5 4 3

Notes on Introductory Combinatorics

George Pólya
Robert E. Tarjan
Donald R. Woods

In the winter of 1978, Professors George Pólya and Robert Tarjan teamed up at Stanford University to teach a course titled "Introduction to Combinatorics". This book consists primarily of the class notes and related material produced by Donald Woods as teaching assistant for the course.

Among the topics covered in the notes are elementary subjects such as combinations and permutations, mathematical tools such as generating functions and Pólya's Theory of Counting, and specific problems such as Ramsey Theory, matchings, and Hamiltonian and Eulerian paths.

PREFACE

In the winter of 1978, Professor George Pólya and I jointly taught Stanford University's introductory combinatorics course. This was a great opportunity for me, as I had known of Professor Pólya since having read his classic book, *How to Solve It*, as a teenager. Working with Pólya, who was over ninety years old at the time, was every bit as rewarding as I had hoped it would be. His creativity, intelligence, warmth and generosity of spirit, and wonderful gift for teaching continue to be an inspiration to me.

Combinatorics is one of the branches of mathematics that play a crucial role in computer science, since digital computers manipulate discrete, finite objects. Combinatorics impinges on computing in two ways. First, the properties of graphs and other combinatorial objects lead directly to algorithms for solving graph-theoretic problems, which have widespread application in non-numerical as well as in numerical computing. Second, combinatorial methods provide many analytical tools that can be used for determining the worst-case and expected performance of computer algorithms. A knowledge of combinatorics will serve the computer scientist well.

Combinatorics can be classified into three types: enumerative, existential, and constructive. Enumerative combinatorics deals with the counting of combinatorial objects. Existential combinatorics studies the existence or nonexistence of combinatorial configurations. Constructive combinatorics deals with methods for actually finding specific configurations (as opposed to merely demonstrating their existence theoretically). The first two-thirds of our course, taught by Professor Pólya, dealt with enumerative combinatorics, including combinations, generating functions, the principle of inclusion and exclusion, Stirling numbers, and Pólya's own theory of counting. The last third of the course, taught by me, covered existential combinatorics, with an emphasis on algorithmic graph theory, and included matching, network flow, Hamiltonian and Eulerian paths, and planar graphs.

Donald Woods, our teaching assistant, was not only invaluable in helping us give the course but also was able to prepare readable and comprehensive course notes, which he has edited to form the present book. Don did a masterful job in making sense out of our

ramblings and adding observations and references of his own. Were
I to teach the course again these notes would be indispensable. I
hope you will enjoy them.

Robert E. Tarjan
Murray Hill, New Jersey
May 3, 1983

Table of Contents

1 | Introduction

For the most part the notes that comprise this text differ only slightly from those provided to the students during the course. The notes have been merged into a single paper, a few sections have been made more detailed, and various corrigenda have been incorporated. The midterm and final examinations are included in their proper chronological places within the text (chapters 8 and 15), together with the solutions. The only information omitted from this book is that regarding the mechanics of the course—office hours, grading criteria, etc. Homework assignments are included, as they often led to further discussion in the notes. Lecture dates are included to give a feel for the pace at which material was covered, though it should be noted that much of the material in the notes was not actually presented in the lectures, being instead drawn from supplementary notes provided by the instructors or supplied by the author as the notes were written.

A brief word of explanation regarding the dual instructorship of the course: Professor Pólya taught the first two-thirds of the course, reflected in chapters 2 through 7 of this book. Professor Tarjan taught the remainder of the course, as covered in chapters 9 through 14.

Though there was no formal text for the course, a number of books were made available for reference. These books, along with additional texts used by the author in preparing the notes, are listed in the bibliography. The [bracketed] abbreviations given there will be used when referring to one of Pólya's books; the other texts will be referred to by their authors. Though all of the books contain relevant material, not all are specifically referenced in this book. In particular, all mentions of [Harary] refer to *Graph Theory* and not to *A Seminar on Graph Theory*.

The author would like to thank Chris Van Wyk and Jim Boyce for their assistance in preparing the manuscript, Don Knuth and Ron Graham for encouraging the publication of the notes and, of course, Professors Pólya and Tarjan for providing ample source material.

G. Pólya et al., *Notes on Introductory Combinatorics*, Modern Birkhäuser Classics,
DOI 10.1007/978-0-8176-4953-1_1, © Birkhäuser Boston, a part of Springer
Science+Business Media, LLC 2010

January 5. In his first lecture, Pólya discussed in general terms what combinatorics is about: The study of counting various combinations or configurations. He started with a problem based on the mystical sign known, appropriately, as an "abracadabra".

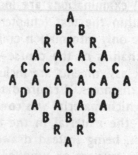

The question is, how many different ways are there to spell out "abracadabra", always going from one letter to an adjacent letter? Due to the way some letters (especially C and D) are found only in certain rows, it turns out the only ways to spell "abracadabra" start with the topmost 'A' and zig-zag down to the bottommost 'A'. If we think of the letters as points, then any spelling of "abracadabra" specifies a sequence of points forming a crooked line from the top to the bottom. One such line is shown below.

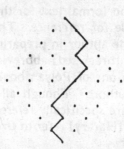

You can also think of this problem in terms of a network of streets in a city where all blocks are the same size. Then the problem becomes one of computing how many ways there are of getting from the northern corner to the southern corner in the minimum number (10) of blocks. (That 10 is the minimum can be seen from the fact that each block, in addition to taking us either east or west, takes us

G. Pólya et al., *Notes on Introductory Combinatorics*, Modern Birkhäuser Classics,
DOI 10.1007/978-0-8176-4953-1_2, © Birkhäuser Boston, a part of Springer
Science+Business Media, LLC 2010

southward one-tenth the t tal southward distance between the two corners.)

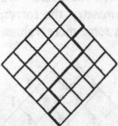

It was decided empirically (*i.e.*, by taking a vote) that there were more than 100 paths, but there was disagreement over whether there were more than 1000, so Pólya proceeded to approach the problem by more formal methods. He began by emphasising an important maxim which you should always consider when working on *any* problem: "*If you cannot solve the proposed problem, solve first a suitable related problem.*" In this instance, the related problem is that of computing how many different paths there are from the northern corner to various other corners, still restricting ourselves to travelling only southeast and southwest. For starters, there is only one path to each of the corners on the northeast edge, namely the path consisting of travelling always southeast and never southwest. Similarly, there is only one path to each of the corners on the northwest edge. We note these values by writing them next to the corners involved.

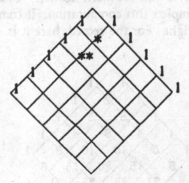

Now what about the corner marked with a *? You could get there by going one block southeast followed by one block southwest, or by going first southwest and then southeast. Similarly, to get to the corner marked **, you could go southeast, then southwest twice, or you could go southwest, then southeast, then southwest, or you could go southwest twice and then southeast. Moving down the diagonal in the manner and, by symmetry, the corresponding diagonal on the eastern side, we can fill in some more values.

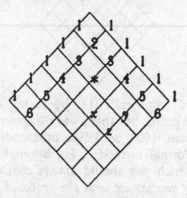

Had we tried to go much further like this, it would probably have gotten rather tiresome, so instead we came up with a general observation regarding an arbitrary corner, such as the one marked z above. If we know that there are x ways to get to the corner just northwest of z, and y ways to get to the corner northeast of z, then there are x+y ways to get to z, since to get there we must first get to either x or y, after which there's only one way to continue on to z. For instance, there are 3+3=6 paths to the corner marked *. This general rule provides us with an easy method to finish computing the number of paths to the southern corner. The first homework assignment was to complete this computation. It comes as no surprise that everyone got it right. For the record, here it is.

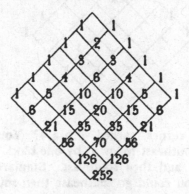

The numbers we've been computing are known as <u>binomial coefficients,</u> for reasons we'll get to eventually. The arrangement of the numbers, when cut off by any horizontal line so as to form a triangular pattern, is known as <u>Pascal's triangle</u>. (Pascal referred to it as "the arithmetical triangle".) The numbers are uniquely defined by the boundary condition (the 1's along the edges) together with the <u>recursion formula</u> (each number not on the edge is the sum of the two above it). In addition to this recursion formula, which defines each number in terms of earlier ones, there is another way to look at the situation. Here's a small chunk of the street network we've been working with:

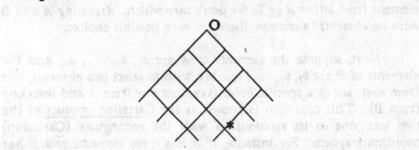

Suppose we want to know the number of different paths (of minimum length) from the origin O to the starred corner. Each such path must consist of 5 blocks, of which exactly 3 go to the right (as seen from above). If we specify which 3 of the 5 blocks will go to the right, we uniquely specify the path. For instance, if we choose the 1st, 4th, and 5th blocks, we get this path:

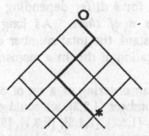

Conversely, each path from O to * specifies a unique set of 3 blocks that go to the right. So the number of paths is the same as the number of ways of choosing 3 blocks out of the total 5. Euler's notation for this sort of thing is $\binom{5}{3}$ or, in general, $\binom{n}{r}$, denoting the number of ways of choosing a subset of size r from a set of size n.

This is usually read "n-choose-r". (Another name often heard to describe this value, but which recently has fallen out of favor, is that used by Jacob Bernoulli: the combinations of n elements taken r at a time.) Computing this value is the first problem of combinatorics.

Next we come to some basic rules for working with multiple sets. The rules are fairly simple (as basic rules are wont to be), but are nevertheless very important (again as basic rules are wont to be). First off, suppose that out of a set of possibilities, A, it is possible to choose any one of m different elements. From another set, B, it is possible to choose any one of n elements. We wish to select an element from either A or B; we don't care which. *Assuming A and B have no elements in common*, there are $m+n$ possible choices.

Next, suppose the elements of A are $a_1, a_2, \ldots, a_m$, and the elements of B are $b_1, b_2, \ldots, b_n$. We wish to select two elements, one from each set, in a specific order (say, first one from A and then one from B). This operation is known as the Cartesian product of the two sets, due to its relationship with the rectangular (Cartesian) coordinate system. For instance, if A has three elements and B has two, there are six possible pairs: (a_1,b_1), (a_1,b_2), (a_2,b_1), (a_2,b_2), (a_3,b_1), and (a_3,b_2). In general, there are $m \cdot n$ possibilities.

Finally, take a more general case of the Cartesian product. Suppose that, having chosen a_1, we then have a choice among a set of elements $b_{11}, b_{12}, \ldots, b_{1n}$. If we start by choosing a_2, we then have a choice from a *different* set: $b_{21}, b_{22}, \ldots, b_{2n}$, and so on. In general, the possibilities for b differ depending upon our choice for a, *but there are always n of them*. As long as the number of possibilities for b is constant, the total number of pairs (a_i,b_j) is still $m \cdot n$. We'll see an application of this in a moment.

A permutation is an ordering of a set of objects. For instance, given the set of three numbers $\{1,2,3\}$, we could order them in any of 6 different ways: $\{1,2,3\}$, $\{1,3,2\}$, $\{2,1,3\}$, $\{2,3,1\}$, $\{3,1,2\}$, or $\{3,2,1\}$. The number of different permutations of n elements is denoted by P_n. Hence $P_3 = 6$. We also see fairly easily that $P_1 = 1$ and $P_2 = 2$. At this point Pólya stated another important maxim: *"The beginning of most discoveries is to recognise a pattern."* There is a pattern to the three numbers we've got so far; to make it more apparent, we can rewrite them as follows:

$$P_1 = 1 = 1$$
$$P_2 = 2 = 1 \cdot 2$$
$$P_3 = 6 = 1 \cdot 2 \cdot 3.$$

We conjecture that $P_n = 1 \cdot 2 \cdot 3 \cdot \ldots \cdot n$. This product is called $\underline{n\ factorial}$ and is usually written "$n!$". Now we need to prove our conjecture. Well, *suppose* it's true that $P_n = n!$. Then what would P_{n+1} be? It is the number of ways of ordering $n+1$ objects. The $n+1^{\text{st}}$ object could be in any one of $n+1$ positions. Whichever of these positions we choose, the remaining n objects can be ordered in any of P_n ways. Using the generalisation of the Cartesian product rule, we conclude that the total number of ways we can order $n+1$ objects is $(n+1) \cdot P_n$. Therefore, if $P_n = 1 \cdot 2 \cdot 3 \cdot \ldots \cdot n$, then $P_{n+1} = 1 \cdot 2 \cdot 3 \cdot \ldots \cdot n \cdot (n+1) = (n+1)!$. But we know that $P_3 = 3!$, so taking $n=3$ we conclude that $P_4 = 4!$. Knowing this, we can take $n=4$ and conclude that $P_5 = 5!$, and so on. For any finite n, we can prove that $P_n = n!$ by starting at P_3 and chugging away for a while. This method of proof, which Pólya describes as "a diabolic way of proving things", is called $\underline{\text{mathematical induction}}$. It is extremely useful since it saves you from having to figure out the formula you're proving. If you can make a "lucky guess" as to what the answer is, you may be able to prove it by induction.

January 10. Pólya began the lecture by reviewing the material from the previous lecture. In doing so he brought out some points that hadn't been explicitly stated before. In particular, there's the formal definition of the binomial coefficients:

Boundary condition: $\binom{n}{0} = \binom{n}{n} = 1$

Recursion: $\binom{n+1}{r} = \binom{n}{r-1} + \binom{n}{r}$.

[n and r integers, $0 < r < n+1$]

Similarly, P_n can be defined by boundary conditions and recursion:

Boundary condition: $P_1 = 1! = 1$

Recursion: $P_n = n! = nP_{n-1}$.

If we apply this recursion formula with $n=1$, we find that $P_1 = 1 \cdot P_0$. Hence we define $P_0 = 0! = 1$.

From here, we move on to look at something Pólya called a "variation", a word you may immediately forget. It is defined as follows. Given a set of n objects, we wish to choose r of them <u>in</u> <u>some order</u>. That is, choosing the first object and then the second would be considered different from choosing the second and then the first. How many such variations are there? One approach is to start by choosing some object to be the first one selected. There are n choices. For each choice, there are $n-1$ choices for the second object. Thus, by the product rule, there are $n(n-1)$ choices for the first two objects together. For each such pair, there are $(n-2)$ objects remaining from which to choose the third object. So there are $n(n-1)(n-2)$ choices for the first three objects. Continuing in this manner, we find that there are $n(n-1)(n-2) \ldots (n-r+1)$ variations.

We can often learn something by solving a problem in two different ways, so here's a second approach. We first choose the subset of r objects from among the n. We know there are $\binom{n}{r}$ ways to do this. We then choose the ordering for the r objects. We know how many ways there are to do this, too; it's P_r. So there are $\binom{n}{r} P_r$ variations. But this answer must be the same as the one we got the other way. Therefore $\binom{n}{r} \cdot P_r = n(n-1)(n-2) \ldots (n-r+1)$. So we have learned something new:

$$\binom{n}{r} = \frac{n(n-1)(n-2) \ldots (n-r+1)}{r!}$$

$$= \frac{n(n-1)(n-2) \ldots (n-r+1)}{1 \cdot 2 \cdot 3 \cdot \ldots \cdot r}.$$

(Note that, in the second form, the sum of 'corresponding' terms in the numerator and denominator is always $n+1$; this can be a useful mnemonic for remembering what the last term in the numerator is.) For example, the number that we computed for the first homework assignment is $\binom{10}{5}$, which by this formula is $(10 \cdot 9 \cdot 8 \cdot 7 \cdot 6)/(1 \cdot 2 \cdot 3 \cdot 4 \cdot 5) = (10 \cdot 9 \cdot 7 \cdot 6)/(1 \cdot 3 \cdot 5) = (2 \cdot 9 \cdot 7 \cdot 6)/3 = 2 \cdot 9 \cdot 7 \cdot 2 = 252$. It's always a good idea to test out a formula on some special cases where we already know the answer, so let's look at $\binom{n}{n}$ and $\binom{n}{0}$. We have

$$\binom{n}{n} = \frac{n(n-1)(n-2) \ldots 1}{1 \cdot 2 \cdot 3 \cdot \ldots \cdot n},$$

which, since the numerator and denominator have all the same factors, albeit in different orders, indeed equals 1. $\binom{n}{0}$, however, poses a bit of a problem, since the numerator has no factors. By defining the product of zero factors to be equal to 1 (just as $0! = 1$) we find that $\binom{n}{0} = 1$ as expected.

Another way we can get this explicit form for the binomial coefficients is by using mathematical induction. We assume it's true for small n (we can check this by hand) and then show that, if it's true for n, it's true for $n+1$. The first problem on the second homework assignment was to carry out this proof. Here it is: We assume that, for some value of n,

$$\binom{n}{r} = \frac{n(n-1)(n-2)\ldots(n-r+1)}{1\cdot 2\cdot 3\cdot\ldots\cdot r}$$

for all values of r. Substituting $r-1$ for r, we find

$$\binom{n}{r-1} = \frac{n(n-1)(n-2)\ldots(n-r+2)}{1\cdot 2\cdot 3\cdot\ldots\cdot(r-1)}.$$

By the definition of the binomial coefficients, we know that

$$\binom{n+1}{r} = \binom{n}{r-1} + \binom{n}{r}$$

$$= \frac{n(n-1)(n-2)\ldots(n-r+2)}{1\cdot 2\cdot 3\cdot\ldots\cdot(r-1)} + \frac{n(n-1)(n-2)\ldots(n-r+1)}{1\cdot 2\cdot 3\cdot\ldots\cdot r}$$

$$= \frac{n(n-1)(n-2)\ldots(n-r+2)\cdot r}{1\cdot 2\cdot 3\cdot\ldots\cdot(r-1)\cdot r} + \frac{n(n-1)(n-2)\ldots(n-r+1)}{1\cdot 2\cdot 3\cdot\ldots\cdot r}$$

$$= \frac{n(n-1)(n-2)\ldots(n-r+2)\cdot(r + n-r+1)}{1\cdot 2\cdot 3\cdot\ldots\cdot(r-1)\cdot r}$$

$$= \frac{(n+1)n(n-1)(n-2)\ldots(n-r+2)}{1\cdot 2\cdot 3\cdot\ldots\cdot r},$$

which is the formula we're trying to prove (with $n+1$ substituted for n). Hence, if the formula is true for n, it's true for $n+1$. This, combined with the fact that it's true for $n=1$, means it is true for all finite n. (Actually, there's a minor flaw in this proof. To wit, the recursion formula cannot be used to compute $\binom{n}{n}$ or $\binom{n}{0}$, since it would involve coefficients outside the range $0 \le r \le n$. However, we've already shown separately that these two special cases satisfy the formula, so we're all right.)

A more compact way to write the formula for the binomial coefficient can be derived by multiplying both the numerator and denominator by the factors $(n-r)$, $(n-r-1)$, and so on down to 1.

$$\binom{n}{r} = \frac{n(n-1)(n-2)\ldots(n-r+1)\cdot(n-r)(n-r-1)\ldots 2\cdot 1}{(1\cdot 2\cdot 3\cdot\ldots\cdot r)(1\cdot 2\cdot\ldots\cdot(n-r-1)\cdot(n-r))}$$

$$= \frac{n!}{r!(n-r)!}$$

Notice that, based on this formula, it is immediately apparent that $\binom{n}{r} = \binom{n}{n-r}$. This was to be expected, since by the method of its construction Pascal's triangle is clearly symmetric.

Next, we consider n houses. They are built identically, because it's easier that way. But then, to make them look different, they are painted different colors: r of them are painted red, s of them yellow, and the remaining t of them green. In how many ways can we assign the colors to the houses? We first choose which houses will be painted red; there are $\binom{n}{r}$ ways to make this choice. Whatever choice we make, there are $n-r$ houses left, of which we choose s to be painted yellow; there are $\binom{n-r}{s}$ ways to do this. At this point we have no choices left to make, since all the rest must be green (that is, $r+s+t=n$). So what do we have? By the product rule, there are $\binom{n}{r}\binom{n-r}{s}$ ways to paint the houses. Using the formula we worked out a moment ago, we find

$$\binom{n}{r}\binom{n-r}{s} = \frac{n!}{r!(n-r)!}\cdot\frac{(n-r)!}{s!(n-r-s)!}.$$

But $n-r-s=t$, and the $(n-r)!$ factors cancel, leaving us with

$$\frac{n!}{r!s!t!},$$

which is, fortunately, symmetric with respect to r, s, and t. (The alternative to its being symmetric would be for it to be wrong, since the original problem was symmetric.) This sort of formula is called a __multinomial coefficient__.

3 | Generating Functions

Generating functions are a general mathematical tool developed by de Moivre, Stirling, and Euler in the 18th century, and are used often in combinatorics. As usual, we start by taking a concrete example: In how many ways can you make change for a dollar? We'll assume that we're dealing with only five types of coins—pennies, nickels, dimes, quarters, and half dollars.

We first consider how many pennies to use. We could use one, or two, or three, etc., and of course we could use none. We can show these choices pictorially:

Similarly, we have an infinite number of choices as to how many nickels we use (although for almost all such choices we'll have more than a dollar already), and how many dimes, and so on:

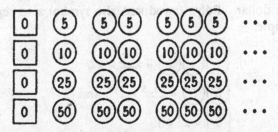

In giving change for a dollar, or for any other amount, we are effectively choosing exactly one 'heap' from each of the five rows. Within each row, we'll represent the fact that we are choosing a single element by writing the row as a summation:

$$\boxed{0} + ① + ①① + ①①① + \cdots$$

Next, we represent the combining of the choices from the various rows by writing the product of the rows (the reason for all this will be seen shortly).

G. Pólya et al., *Notes on Introductory Combinatorics*, Modern Birkhäuser Classics, DOI 10.1007/978-0-8176-4953-1_3, © Birkhäuser Boston, a part of Springer Science+Business Media, LLC 2010

$$\left(\boxed{0} + ① + ①① + ①①① + \cdots \right)$$
$$\times \left(\boxed{0} + ⑤ + ⑤⑤ + ⑤⑤⑤ + \cdots \right)$$
$$\times \left(\boxed{0} + ⑩ + ⑩⑩ + ⑩⑩⑩ + \cdots \right)$$
$$\times \left(\boxed{0} + ㉕ + ㉕㉕ + ㉕㉕㉕ + \cdots \right)$$
$$\times \left(\boxed{0} + ㊿ + ㊿㊿ + ㊿㊿㊿ + \cdots \right)$$

Now, why did we do this? Well, if we look at the infinite product we've created, we find that each term in the product is the product of five terms, one from each of the sums. Thus, each term of the product corresponds to a different combination of coins, and if we look at *all* the terms of the product, we'll find they include *all* such combinations.

But we don't *want* all combinations; we just want the ones that add up to a dollar. Pólya introduced the symbol x to represent 1¢. So, for example,

$$①①① = xxx = x^3,$$
$$⑤⑤ = x^5 x^5 = x^{10}, \text{ and}$$
$$\boxed{0} = x^0 = 1.$$

Our product can now be written more mathematically as follows:

$$(1 + x + x^2 + x^3 + \cdots)$$
$$\cdot (1 + x^5 + x^{10} + x^{15} + \cdots)$$
$$\cdot (1 + x^{10} + x^{20} + x^{30} + \cdots)$$
$$\cdot (1 + x^{25} + x^{50} + x^{75} + \cdots)$$
$$\cdot (1 + x^{50} + x^{100} + x^{150} + \cdots)$$

For example, one of the terms in the product will be $x^3 \cdot x^5 \cdot x^{20} \cdot 1 \cdot x^{50}$, which corresponds to the combination of coins that consists of 3 pennies, 1 nickel, 2 dimes, no quarters, and 1 half dollar. When the five terms, one from each infinite sum, are multiplied, the exponents add; this is just what we want, because it means the exponent (in our example it's 78) is the total value of the selected coins. So for each combination of coins totalling one dollar, there will be a term in the product with an exponent of 100. If we combine terms that have the same exponent, we get something of the form

$$1 + E_1 x + E_2 x^2 + \cdots + E_{100} x^{100} + \cdots.$$

January 12. All we need to do is find the coefficient E_{100}. But how do we do that? We could try multiplying out the infinite product, but this would probably take a while. Instead we use what we know about series, and in particular about geometric series.

Consider a typical geometric series: $1 + x + x^2 + x^3 + \cdots.$ What does this series sum to? Pólya claimed it was obvious: the sum is S. That doesn't sound like it helps much but, having given it a name, we can manipulate it mathematically. In particular, we can multiply S by $(1-x)$.

$$S(1-x) = 1 + x + x^2 + x^3 + \cdots$$
$$\qquad\qquad - x - x^2 - x^3 - x^4 - \cdots$$
$$= 1$$

So $S = 1/(1-x)$. Similarly, $1 + x^5 + x^{10} + x^{15} + \cdots = 1/(1-x^5)$. Our infinite product thus simplifies to the somewhat more compact form

$$\frac{1}{(1-x)(1-x^5)(1-x^{10})(1-x^{25})(1-x^{50})},$$

which we can turn back into a series in powers of x, even though we don't yet know the coefficients numerically, as

$$\sum_{n=0}^{\infty} E_n x^n.$$

Such a summation, in either form, is called the <u>generating function</u> for the sequence $E_0, E_1, E_2, \ldots.$

So far so good, but we don't appear to be any closer to computing E_{100} than we were before. Once again we'll try first solving an easier related problem. In fact, we'll set up a sequence of problems leading to the one we're interested in.

$$\frac{1}{(1-x)} = \sum_{n=0}^{\infty} A_n x^n$$

$$\frac{1}{(1-x)(1-x^5)} = \sum_{n=0}^{\infty} B_n x^n$$

$$\frac{1}{(1-x)(1-x^5)(1-x^{10})} = \sum_{n=0}^{\infty} C_n x^n$$

$$\frac{1}{(1-x)(1-x^5)(1-x^{10})(1-x^{25})} = \sum_{n=0}^{\infty} D_n x^n$$

$$\frac{1}{(1-x)(1-x^5)(1-x^{10})(1-x^{25})(1-x^{50})} = \sum_{n=0}^{\infty} E_n x^n$$

We already know that $A_n = 1$ for all $n \geq 0$. What about B_n? We take the second equation above and multiply both sides by $1-x^5$. The left side becomes $1/(1-x)$, which is series A.

$$\sum_{n=0}^{\infty} A_n x^n = (1-x^5)(\sum_{n=0}^{\infty} B_n x^n)$$

$$= (\sum_{n=0}^{\infty} B_n x^n) - (\sum_{n=0}^{\infty} B_n x^{n+5})$$

What does it mean for these two sums to be equal? Since they must be equal for all values of x, it means that the coefficients of x^n must be equal for all n. On the left side the coefficient is simply A_n. On the right side, the first summation contributes a term of $B_n x^n$, and the second summation contributes $-B_{n-5} x^n$ (coming from multiplying $(-x^5)$ by $B_{n-5} x^{n-5}$). Therefore

$$A_n = B_n - B_{n-5}$$

or, rearranging things,

$$B_n = A_n + B_{n-5}.$$

By the same reasoning we can also find that

$$C_n = B_n + C_{n-10}$$
$$D_n = C_n + D_{n-25}$$
$$E_n = D_n + E_{n-50}.$$

For boundary conditions, we know that none of the series has any terms with x^n for $n < 0$, and so $A_n = B_n = C_n = D_n = E_n = 0$ for $n < 0$. We also know that $A_n = 1$ for all $n \geq 0$. Armed with this information, we can compute B_n for $n \geq 0$, making use of the recursion formula we've just worked out. Thus, $B_0 = A_0 + B_{-5} = 1 + 0 = 1$, $B_1 = A_1 + B_{-4} = 1 + 0 = 1, \ldots$, $B_5 = A_5 + B_0 = 1 + 1 = 2$, etc. Once we've worked out some of the B's, we can start computing C's, and so on. Even so, working all the way out to E_{100} by hand could be time-consuming, though it wouldn't take long using a computer. But we can save a lot of effort by observing that we don't need *all* the intermediate numbers. To compute E_{100} we need to know E_{50}, and to compute that we need E_0. We also need D_{100}, D_{50}, and D_0. To compute *those*, we also need to know D_{75} and D_{25}, and so on. So if we plan ahead a little bit, we can compute only those elements we actually need.

Pólya demonstrated the process by beginning to fill in a table with $n = 0, 10, 20, \ldots, 100$. The second problem on the second homework assignment was to finish the computation and find E_{100}. (Pólya also provided as a hint that E_{50} happens to be 50.) He failed to point out that some intermediate multiples of 5 would also be necessary, but everyone seemed to figure that out anyway. The following table shows the minimum number of entries that need to be filled in to get the final answer of 292. (Some of the entries, such as B_{85} and B_{95}, could be left out by observing (and proving) some simple patterns, such as $B_n = A_n + B_{n-5} = A_n + (A_{n-5} + B_{n-10}) = 2 + B_{n-10}$ for $n \geq 10$, but we'll work them out anyway.)

n	0	5	10	15	20	25	30	35	40	45	50	55	60	65	70	75	80	85	90	95	100
B_n	1	2	3	4	5	6	7	8	9	10	11	12	13	14	15	16	17	18	19	20	21
C_n	1	2	4	6	9	12	16	20	25	30	36	42	49	56	64	72	81		100		121
D_n	1			13							49					121					242
E_n	1										50										292

Here is a summary of some of the more useful rules regarding generating functions. Suppose we have two generating functions:

$$g(x) = a_0 + a_1 x + a_2 x^2 + \cdots + a_n x^n + \cdots$$
$$h(x) = b_0 + b_1 x + b_2 x^2 + \cdots + b_n x^n + \cdots$$

Then:

[1] $g(x) = h(x) \iff a_0 = b_0,\ a_1 = b_1,\ a_2 = b_2$, etc.

[2] $g(x) + h(x) = (a_0 + b_0) + (a_1 + b_1)x + \cdots + (a_n + b_n)x^n + \cdots$

[3] $g(x) \cdot h(x) = (a_0 b_0 + a_0 b_1 x + a_0 b_2 x^2 + \cdots)$
$$+ (a_1 b_0 x + a_1 b_1 x^2 + a_1 b_2 x^3 + \cdots)$$
$$+ (a_2 b_0 x^2 + a_2 b_1 x^3 + a_2 b_2 x^4 + \cdots)$$
$$+ \cdots$$
$$= (a_0 b_0) + (a_0 b_1 + a_1 b_0)x + (a_0 b_2 + a_1 b_1 + a_2 b_0)x^2 + \cdots$$
$$= c_0 + c_1 x + c_2 x^2 + \cdots + c_n x^n + \cdots ,$$

where c_n is defined as $= a_0 b_n + a_1 b_{n-1} + a_2 b_{n-2} + \cdots + a_n b_0$.

January 17. Generating functions worked well on the problem of changing a dollar; let's try applying them elsewhere. We'll start with something we already know about—binomial coefficients. Suppose we have a set of n objects $\{x_1, x_2, x_3, \ldots, x_n\}$, and we wish to choose a subset of r objects. We either choose x_1 or we don't. As before, we'll represent this choice by a sum of the possibilities, $(x_1{}^0 + x_1{}^1)$, or simply $(1 + x_1)$. Similarly we have $(1 + x_2)$, $(1 + x_3)$, and so on. We again represent the combination of choices by a product:

$$(1 + x_1)(1 + x_2)(1 + x_3) \ldots (1 + x_n).$$

Each term of this product constitutes a selection of exactly one term from each of the n sums, which corresponds to a selection of some number (not necessarily r) of objects from the original set. For instance, if we choose the x_1 from the first sum and the 1 from each of the others, we get the term x_1. The product comes out to

$$1 + x_1 + x_2 + x_3 + \cdots + x_n$$
$$+ x_1 x_2 + x_1 x_3 + x_2 x_3 + \cdots + x_{n-1} x_n$$
$$+ x_1 x_2 x_3 + x_1 x_2 x_4 + \cdots + x_{n-2} x_{n-1} x_n$$
$$\vdots$$
$$+ x_1 x_2 x_3 \ldots x_n.$$

The number of ways of choosing, say, two x's is the number of terms that contain exactly two x's. So let all the x_i be equal; that is, let

$x_1 = x_2 = x_3 = \cdots = x_n = x$. Then

$$(1+x_1)(1+x_2)(1+x_3) \ldots (1+x_n) = (1+x)^n = a_0 + a_1 x + a_2 x^2 + \cdots + a_n x^n.$$

It's clear that a_0 is 1; what about a_1? It is equal to the number of terms in the product that contain exactly one x, which is therefore the number of different subsets of size 1. Hence $a_1 = \binom{n}{1}$. Similarly, $a_2 =$ the number of different subsets of size $2 = \binom{n}{2}$, and so on. For that matter, $a_0 = 1 = \binom{n}{0}$. We can summarise all this in one handy equation:

$$(1+x)^n = \sum_{k=0}^{n} \binom{n}{k} x^k.$$

This is called the "binomial formula" (because $1+x$ is the "basic" polynomial of two terms); hence the name "binomial coefficients".

Pólya next brought up a third maxim: *If you have a general formula, try it out on some special cases.* One special case is $x = 1$. This gives us

$$2^n = \binom{n}{0} + \binom{n}{1} + \binom{n}{2} + \cdots + \binom{n}{n},$$

which is the number of subsets of *all* sizes from a set of size n. This checks, since for each object we have two choices—either it is in the subset or it isn't. We have n such choices, and by the product rule the total number of possibilities is therefore the product of n 2's.

Another interesting special case, which didn't come up in the lecture, is that of $x = -1$:

$$0^n = \binom{n}{0} - \binom{n}{1} + \binom{n}{2} - \cdots + (-1)^n \binom{n}{n}.$$

That this sum should be zero is obvious when n is odd, due to the symmetry of Pascal's triangle. When n is even, however, the above result is less obvious, so this identity is worth noting. Note in particular that, substituting the value $n=0$, we can deduce 0^0 must equal 1.

Let's consider the combinations of n objects <u>with repetition</u> taken r at a time, which we'll denote by $R_r^{(n)}$. We can also think of it as having n <u>kinds</u> of objects, with an unlimited supply of each, from which we wish to select r objects. Let's find the generating function for this.

Just as when we were looking at ways to break a dollar, we can have no x_1's, or one x_1, or two, or three, etc., and we'll write this as the sum $1+x_1+x_1x_1+x_1x_1x_1+\cdots$, and similarly for each x in the set. We take the product,

$$(1+x_1+x_1x_1+x_1x_1x_1+\cdots)\cdot(1+x_2+x_2x_2+x_2x_2x_2+\cdots)\cdot\ldots$$
$$\cdot(1+x_n+x_nx_n+x_nx_nx_n+\cdots),$$

so that, as usual, each term of the product corresponds to one of the possible selections. We're interested in the selections that include exactly r x's (not necessarily different x's), and we don't want to distinguish among the x's, so we let $x_1 = x_2 = \cdots = x_n = x$, and get

$$(1+x+x^2+x^3+\cdots)^n.$$

Each term that selects exactly r x's contributes 1 to the coefficient of x^r, so that's the coefficient we want. Well, we already know what the geometric series sums to, so what we've got is

$$\left(\frac{1}{1-x}\right)^n = 1 + \cdots + R_r^{(n)}x^r + \cdots.$$

Let's digress for a moment and examine a useful generalisation of binomial coefficients. Newton defined $\binom{\alpha}{r}$ for non-integer α as

$$\binom{\alpha}{r} = \frac{\alpha(\alpha-1)(\alpha-2)\ldots(\alpha-r+1)}{1\cdot2\cdot3\cdot\ldots\cdot r}$$

and claimed that, if $|x|<1$,

$$(1+x)^\alpha = \sum_{k=0}^{\infty}\binom{\alpha}{k}x^k$$

for *all* α. Newton didn't actually prove this (rigorous proofs were not recognised as being necessary in his day), but Gauss proved it in 1812. We won't bother to show the proof here. Using this result, we find

$$(1-x)^{-n} = \sum_{k=0}^{\infty}\binom{-n}{k}(-x)^k.$$

Since $R_r^{(n)}$ is the coefficient of x^r in this sum, we find

$$R_r^{(n)} = \binom{-n}{r} \cdot (-1)^r$$

$$= \frac{(-n)(-n-1)(-n-2)\ldots(-n-r+1)}{1 \cdot 2 \cdot 3 \cdot \ldots \cdot r}(-1)^r.$$

Combining the r factors of -1 with the r terms in the numerator, we find that

$$R_r^{(n)} = \frac{n(n+1)(n+2)\ldots(n+r-1)}{1 \cdot 2 \cdot 3 \cdot \ldots \cdot r}$$

$$= \binom{n+r-1}{r}.$$

a perfectly ordinary binomial coefficient.

As usual, we'll try to confirm this by proving it another way. Suppose we have n kinds of objects, $x_1, x_2, \ldots, x_n$. We select r objects, and set them down in order of increasing subscript, with "separation points" every time we come to a different kind of object. For example:

$$x_1 x_1 x_1 \bullet x_2 x_2 x_2 x_2 \bullet x_3 \bullet \ldots \bullet x_n x_n.$$

Even if we have no x_i for some i, we'll include the separation point:

$$\bullet \bullet x_3 \bullet \bullet x_5 x_5 \bullet \ldots \bullet x_{n-1} x_{n-1} \bullet.$$

Thus we always have r objects plus $n-1$ separation points. Once we choose the positions of the separation points, we have completely determined the set being selected. Everything ahead of the first separation point consists of x_1's, everything between the first and second points is x_2, and so forth. Thus there are as many subsets of size r (with repetition) as there are ways of selecting $n-1$ separation points from among $r+n-1$ possible positions (without repetition). This number is, of course, $\binom{r+n-1}{n-1}$, which by the symmetry of binomial coefficients (i.e., $\binom{m}{t} = \binom{m}{m-t}$) is equal to our earlier answer.

At this point Pólya assigned as "non-obligatory homework" two observations that had nothing at all to do with generating functions, but were simply things that people might find interesting to investigate. Consider Pascal's triangle, shown (in part) below.

```
                              1
                           1     1
                        1     2     1
                     1     3     3     1
                  1     4     6     4     1
               1     5    10    10     5     1
            1     6    15    20    15     6     1
         1     7    21    35    35    21     7     1
      1     8    28    56    70    56    28     8     1
   1     9    36    84   126   126    84    36     9     1
1    10    45   120   210   252   210   120    45    10     1
```

(Please pardon us for not showing all of Pascal's triangle; infinite tables use up too much paper.) The first observation was that, for certain values of n, all of the values in row n (remembering that the top row is row 0) except the first and last elements are divisible by n. For instance, in row 7, we have 7, 21, and 35. Pólya pointed out that this happens whenever n is prime, and suggested it as a topic for further thought. Well, let's think about it.

Why should $\binom{p}{r}$ be divisible by p whenever p is prime and $0 < r < p$? For the answer, look at the formula for $\binom{p}{r}$.

$$\binom{p}{r} = \frac{p(p-1)(p-2)\ldots(p-r+1)}{1 \cdot 2 \cdot 3 \cdot \ldots \cdot r}$$

Note the factor of p in the numerator. Since p is prime, it's not going to cancel out against anything in the denominator unless it's another p. And the denominator won't include a factor of p unless $r = p$. (At the other end, if $r = 0$, the numerator has zero factors, so the factor of p never occurs at all.) Hence for $0 < r < p$, the factor of p cannot be cancelled out, so the resulting value must still have a factor of p. (This is somewhat informal, but a more rigorous proof would require a bit too much number theory.)

An interesting corollary to this is that, for p prime, the sum of all the elements of row p must be 2 greater than a multiple of p, since all the numbers except the two 1's are multiples of p. But we already know what the sum is; it's 2^p. So we've proven that

2^p-2 is a multiple of p for any prime p.

This happens to be a special case of Fermat's theorem (1640) which states that, if p is prime, then a^p-a is a multiple of p for any integer a. The proof of this general theorem again requires more number theory than we want to go into here, but the special case for $a=2$ can be derived directly from combinatorics, as we've just seen.

The second aspect of Pascal's triangle suggested for further study was the number of odd numbers in each row. Starting at the top, we count 1, 2, 2, 4, 2, 4, 4, 8, 2, 4, 4, ... odd numbers per row. What is the pattern here? Several people observed that all of these numbers are powers of two. A few people even determined that the exponent is equal to the number of 1's in the binary representation of the row number, n. That is, row n of the triangle contains exactly $2^{\nu(n)}$ odd numbers, where $\nu(n)$ denotes the number of digits that are 1's in the base two representation of n. For instance, $9_{10} = 1001_2$, which contains two 1's, so row 9 should contain $2^2 = 4$ odd numbers. It does: 1, 9, 9, and 1.

Proving this is somewhat difficult, but for those people who are interested we shall attempt to present a proof that does not make use of any non-obvious results from number theory. If you're not interested, you can skip down to the discussion on polygon dissection (starting on page 25). If you'd rather see a more formal treatment of the problem, look at [Knuth], vol. 1, section 1.2.6, exercise 10e (the answer is in the back of the book).

The proof starts by proving a more general result—a rule for determining whether $\binom{n}{r}$ is odd for any given n and r. First, we'll introduce the notation $\lfloor x \rfloor$ to represent the <u>largest integer less than or equal to x</u>. That is, if $\lfloor x \rfloor = k$, then k is the unique integer satisfying the condition $k \le x < k+1$. Thus, for example, $\lfloor \pi \rfloor = 3$, $\lfloor 7 \rfloor = 7$, and $\lfloor -\pi \rfloor = -4$ (*not* -3). (The notation $[x]$ is also used, but $\lfloor x \rfloor$ is generally preferred these days as being somewhat more mnemonic, particularly when it's used in conjunction with a related function denoted by $\lceil x \rceil$.) Next, we state without proof the following lemma: If an integer k can be written in the form

$$k = \frac{a_1 \cdot a_2 \cdot a_3 \cdot \ldots \cdot a_l}{b_1 \cdot b_2 \cdot b_3 \cdot \ldots \cdot b_l}$$

where all the a's and b's are integers, then: (1) If the a's contain
more factors of two than do the b's (this is *not* the same as the
number of even numbers since, for example, 24 counts as 3 factors of
two), then k is even, since not all of the twos in the numerator will
be cancelled out. (2) If the a's and b's contain the same number of
factors of two, then k is odd, since all of the twos will cancel, leaving
a product of odd numbers in the numerator divided by another
product of odd numbers in the denominator. (3) The b's cannot
contain more factors of two than do the a's, since k could not then be
an integer. We hope that these assertions are intuitively obvious;
rigorous proofs require too much number theory to be included here.

    At this point we're ready to state the main theorem we need
for this analysis. It is this: If n is even and r is odd, then $\binom{n}{r}$ is
even. Otherwise, $\binom{n}{r}$ is even *if and only if* $\binom{\lfloor n/2 \rfloor}{\lfloor r/2 \rfloor}$ is even.

    To prove this, first consider the case where n is even and r is
odd. We have

$$\binom{n}{r} = \frac{n(n-1)(n-2)\ldots(n-r+1)}{1 \cdot 2 \cdot 3 \cdot \ldots \cdot r}$$

$$= \frac{n}{r} \cdot \frac{(n-1)(n-2)\ldots(n-r+1)}{1 \cdot 2 \cdot 3 \cdot \ldots \cdot (r-1)}$$

$$= \frac{n}{r}\binom{n-1}{r-1}.$$

(To justify this completely we must observe that, since r is odd and
$0 \leq r \leq n$, we know $0 \leq r-1 \leq n-1$.) Now let $a_1 = n$, $a_2 = \binom{n-1}{r-1}$, and $b_1 = r$. Since
n is even and r is odd, regardless of what $\binom{n-1}{r-1}$ is (it's an integer;
that's all that counts), the lemma says that $a_1 a_2 / b_1 = \binom{n}{r}$ must be even.

    Next we'll consider the tricky case—n and r both even. We
again start with

$$\binom{n}{r} = \frac{n(n-1)(n-2)\ldots(n-r+1)}{1 \cdot 2 \cdot 3 \cdot \ldots \cdot r}.$$

We observe that the factors $(n-1)$, $(n-3)$, $\ldots$, $(n-r+1)$ are all odd, as
are $1, 3, 5, \ldots, (r-1)$. By the lemma, these factors can be ignored so
far as the even/oddness of $\binom{n}{r}$ is concerned. This leaves $r/2$ terms in
both the numerator and denominator, all of them even:

$$\frac{n(n-2)(n-4)\ldots(n-r+2)}{2\cdot4\cdot6\cdot\ldots\cdot r}.$$

We divide each of the terms by two, which doesn't affect the value of the number since there are the same number of terms in the numerator as in the denominator, and are left with

$$\frac{\tfrac{1}{2}n(\tfrac{1}{2}n-1)(\tfrac{1}{2}n-2)\ldots(\tfrac{1}{2}n-\tfrac{1}{2}r+1)}{1\cdot2\cdot3\cdot\ldots\cdot\tfrac{1}{2}r}.$$

All we've done is throw away some factors that were odd (and therefore contained no factors of two), and divided out an equal number of twos from top and bottom, so by the lemma this new number is even if and only if the original number, $\binom{n}{r}$, was even. But this new number is simply $\binom{n/2}{r/2}$. Since, for k even, $\lfloor k/2\rfloor=(k/2)$, we have proven the theorem for n and r both even.

Next, suppose n and r are both odd. Then by our earlier reasoning we know

$$\binom{n}{r}=\frac{n}{r}\binom{n-1}{r-1}.$$

Since n and r are both odd, the lemma tells us that $\binom{n}{r}$ is even if and only if $\binom{n-1}{r-1}$ is even. But $n-1$ and $r-1$ are both even, so this is the case we've just shown: $\binom{n-1}{r-1}$ is even if and only if $\binom{(n-1)/2}{(r-1)/2}$ is even. Since, for k odd, $\lfloor k/2\rfloor=((k-1)/2)$, we have proven the theorem for n and r both odd.

Finally, suppose n is odd and r is even. Then we know $n\neq r$, so we can multiply $\binom{n}{r}$ by $(n-r)/(n-r)$ to get

$$\binom{n}{r}=\frac{n}{n-r}\cdot\frac{(n-1)(n-2)\ldots(n-r+1)(n-r)}{1\cdot2\cdot3\cdot\ldots\cdot r}$$

$$=\frac{n}{n-r}\binom{n-1}{r}.$$

Since n is odd and $n-r$ is odd, the lemma tells us that $\binom{n}{r}$ is even if and only if $\binom{n-1}{r}$ is even, and the theorem quickly follows as in the other cases. We therefore have finished proving the theorem.

So much for the hard part. Now let's use the theorem to get

our final result. First we make two observations regarding the
binary representation of a number n: (1) n is even if and only if the
last binary digit is a zero; (2) the binary representation of $\lfloor n/2 \rfloor$ is the
same as that of n except the last digit is removed. So by our
theorem, $\binom{n}{r}$ is even if the last binary digits of n and r are 0 and 1,
respectively. If they're not, then we look at $\lfloor n/2 \rfloor$ and $\lfloor r/2 \rfloor$. If *their*
last digits are 0 and 1, respectively, it means that $\binom{\lfloor n/2 \rfloor}{\lfloor r/2 \rfloor}$ is even, so $\binom{n}{r}$
is also even. Otherwise we look next at $\lfloor \lfloor n/2 \rfloor /2 \rfloor$ and $\lfloor \lfloor r/2 \rfloor /2 \rfloor$, and
so forth. But wait a moment; the last binary digits of $\lfloor n/2 \rfloor$ and $\lfloor r/2 \rfloor$
are simply the next-to-last digits of n and r, and the last digits of
$\lfloor \lfloor n/2 \rfloor /2 \rfloor$ and $\lfloor \lfloor r/2 \rfloor /2 \rfloor$ are the third-to-last digits of n and r, etc.
Thus $\binom{n}{r}$ is even if, in *any* digit position, the binary representations
of n and r contain a 0 and 1, respectively. And what if they don't?
In that case we continue discarding digits off the ends of the two
binary representations, and eventually are left with nothing but
zeroes. Since $\binom{0}{0} = 1$, which is odd, $\binom{n}{r}$ must also be odd.

For example, $45_{10} = 101101_2$, and $20_{10} = 10100_2$. Since the
latter contains a 1 in the fifth digit from the right, whereas the
corresponding digit in the first number is a 0, we know that $\binom{45}{20}$ is
even. On the other hand, since $12_{10} = 1100_2$, which contains zeroes
wherever 45 does, we know that $\binom{45}{12}$ is odd. (Feel free to check these
results; <u>we</u> did!)

Okay, we're almost done (finally!). How many numbers in row
n of Pascal's triangle are odd? This is the same as asking how many
numbers r between 0 and n (inclusive) have zeroes wherever n has
zeroes in binary notation. How many such r are there? Well, r's
binary representation has zeroes wherever n's does. Wherever n
contains a 1, however, r can contain either a 1 or a 0. We don't
have to worry about making r larger than n since, even if we put 1's
in *all* such positions, all we get is n itself, and that's the largest we
can possibly make r. So, letting $\nu(n)$ represent the number of 1's in
the binary representation of n, there are exactly $\nu(n)$ binary digits in
r that can be either 0 or 1, and the rest of the digits must be 0. By
the product rule we have $2^{\nu(n)}$ possible values for r, and therefore
there are exactly that many odd numbers in row n of the triangle.

There's a completely different approach to this problem. It
involves looking at the pattern of even and odd numbers. If we
represent an odd number by a • and an even number by a blank,

then the top 64 rows of the triangle look like this:

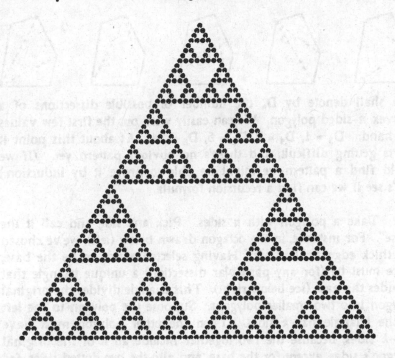

Notice that the pattern in the top 32 rows is duplicated on both sides in the next 32 rows, with nothing but even numbers in between. You might like to give some thought to how you might go about (a) proving this replication pattern in general and (b) using it to prove that row n contains $2^{\nu(n)}$ odd numbers.

Enough already about Pascal's triangle! Let us proceed with the course notes.

January 19. We wish to dissect a convex n-sided polygon into triangles by adding non-intersecting diagonals. In how many ways can this be done? For example, a convex quadrilateral can be dissected into triangles in either of two ways:

A pentagon can be dissected in any of five ways:

We shall denote by D_n the number of possible dissections of a convex n-sided polygon. We can easily work out the first few values by hand: $D_3 = 1$, $D_4 = 2$, $D_5 = 5$, $D_6 = 14$. At about this point it starts getting difficult, and there's no obvious pattern yet. (If we could find a pattern we might be able to prove it by induction.) Let's see if we can find a recursion formula.

Take a polygon with n sides. Pick any side and call it the "base". For instance, in the octagon drawn below (left), we've chosen the thick edge as the base. Having selected a side to be the base, there must be (for any particular dissection) a unique triangle that includes the base (see below right). This triangle divides the original polygon into two smaller polygons. Suppose the polygon to the left of the triangle has k sides. Then the other polygon must have $n+1-k$ sides, because the two together include all n of the original polygon's sides, except for the base, and also the two dotted sides, for a total of $n+1$ sides.

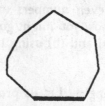 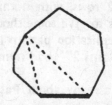

By definition, we know that there are D_k ways to dissect the k-sided polygon. Each such dissection can be combined with any of the D_{n+1-k} possible dissections of the right-hand polygon. By the product rule, the original polygon has $D_k \cdot D_{n+1-k}$ dissections that include this particular triangle at the base. Meanwhile, k can take on various values, depending on what triangle actually includes the base. One particularly strange case that we'll have to watch out for is the one shown on the following page where, once the triangle is removed, we're left with only a single polygon of $n-1$ sides.

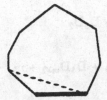

There is, of course, a similar special case with the dotted diagonal going up to the right instead of up to the left. By the rule of sums (way back on page 6 of these notes) we add the configurations for each base triangle and get

$$D_n = D_{n-1} + D_3D_{n-2} + \cdots + D_kD_{n+1-k} + \cdots + D_{n-2}D_3 + D_{n-1}.$$

This would be sufficient if all we wanted to do was program a computer to evaluate D_n, but from an aesthetic standpoint it's not pleasing. Suppose we consider a single edge to be a polygon of 2 sides—up the edge and back along the same edge—and let $D_2 = 1$. Then our equation becomes somewhat more regular.

$$D_n = D_2D_{n-1} + D_3D_{n-2} + \cdots + D_kD_{n+1-k} + \cdots + D_{n-2}D_3 + D_{n-1}D_2$$

Now then, this chapter is supposed to be about generating functions (in spite of all that stuff about Pascal's triangle), so let's make use of them. We'll define

$$g(x) = D_2x^2 + D_3x^3 + \cdots + D_kx^k + \cdots.$$

(Pólya describes this as "putting all the D's in a 'bag'.") Recalling the formula for products of generating functions, we take a look at the square of g.

$$[g(x)]^2 = \left(\sum_{k=2}^{\infty} D_k x^k \right) \cdot \left(\sum_{l=2}^{\infty} D_l x^l \right)$$

$$= \sum_{k=2}^{\infty} \sum_{l=2}^{\infty} D_k D_l x^{k+l}$$

$$= D_2D_2x^4 + (D_2D_3+D_3D_2)x^5 + \cdots$$
$$+ (D_2D_{m-1}+D_3D_{m-2}+\cdots+D_{m-1}D_2)x^{m+1} + \cdots$$

But from our recursion equation, we know

$$D_3 = D_2 D_2$$
$$D_4 = D_2 D_3 + D_3 D_2$$
$$\vdots$$
$$D_m = D_2 D_{m-1} + D_3 D_{m-2} + \cdots + D_{m-1} D_2$$

and therefore

$$[g(x)]^2 = D_3 x^4 + D_4 x^5 + \cdots + D_m x^{m+1} + \cdots$$

$$= -D_2 x^3 + D_2 x^3 + D_3 x^4 + D_4 x^5 + \cdots + D_m x^{m+1} + \cdots.$$

Since $D_2 = 1$, we arrive at

$$[g(x)]^2 = -x^3 + x g(x).$$

Now to solve this quadratic equation. (We'll write g instead of $g(x)$ just to make things more readable.) Pólya's approach was to multiply through by 4, add x^2 to both sides, and move the xg term over, with the following results.

$$4g^2 - 4gx + x^2 = x^2 - 4x^3$$

$$(2g-x)^2 = x^2(1-4x)$$

$$2g - x = \pm x(1-4x)^{\frac{1}{2}}$$

$$g = \tfrac{1}{2}x[1 \pm (1-4x)^{\frac{1}{2}}]$$

What sign do we want to choose for the ±? If we choose '+', then we get into trouble, because the binomial theorem tells us that

$$(1 + z)^{\frac{1}{2}} = 1 + \sum_{k=1}^{\infty} \binom{1/2}{k} z^k,$$

so the leading term of g would be $\tfrac{1}{2}x(1+1) = x$. But we know that g doesn't have any terms ahead of x^2, so we choose '−' instead.

$$g = \tfrac{1}{2}x[1 - (1-4x)^{\frac{1}{2}}]$$

Now let's take a moment to manipulate that square root some more.

$$(1-4x)^{\frac{1}{2}} = 1 + \sum_{k=1}^{\infty} \binom{1/2}{k}(-4x)^k$$

$$= 1 + \sum_{k=1}^{\infty} \frac{\frac{1}{2}(-\frac{1}{2})(-\frac{3}{2}) \ldots (\frac{1}{2}-k+1)}{1 \cdot 2 \cdot 3 \cdot \ldots \cdot k}(-4x)^k$$

$$= 1 - 2x + \sum_{k=2}^{\infty} \frac{(\frac{1}{2})^k 1(-1)(-3) \ldots (1-2k+2)}{1 \cdot 2 \cdot 3 \cdot \ldots \cdot k} 4^k x^k (-1)^k$$

$$= 1 - 2x + \sum_{k=2}^{\infty} \frac{2^k \cdot (-1) \cdot 1 \cdot 3 \cdot 5 \cdot \ldots \cdot (2k-3)}{1 \cdot 2 \cdot 3 \cdot \ldots \cdot k} x^k$$

$$= 1 - 2x - 2 \sum_{k=2}^{\infty} \frac{2 \cdot 6 \cdot 10 \cdot \ldots \cdot (4k-6)}{2 \cdot 3 \cdot 4 \cdot \ldots \cdot k} x^k$$

Hence

$$g = \tfrac{1}{2}x[2x + 2 \sum_{k=2}^{\infty} [(2/2)(6/3)(10/4) \ldots ((4k-6)/k)]x^k]$$

If we let $k+1=n$,

$$g = x^2 + \sum_{n=3}^{\infty} [(2/2)(6/3)(10/4) \ldots ((4n-10)/(n-1))]x^n.$$

Since D_n is the coefficient of x^n in g, we have

$$D_n = (2/2) \cdot (6/3) \cdot (10/4) \cdot (14/5) \cdot \ldots \cdot ((4n-10)/(n-1)),$$

from which we can easily compute D_n for any particular value of n. If we are computing several consecutive D_n, we can take advantage of the observation that

$$D_{n+1} = \frac{4n-6}{n} D_n,$$

and thus compute successive values in roughly constant time. [Note: This problem is in [MPR], vol. 1, page 102, ex. 7, 8, and 9.]

One student included on his homework paper a continuation of the preceding analysis. Rewriting the formula as

$$D_n = \frac{2 \cdot 6 \cdot 10 \cdot \ldots \cdot (4n-10)}{2 \cdot 3 \cdot 4 \cdot \ldots \cdot (n-1)},$$

he decided to look for a more compact equivalent formula. He started by extracting a factor of 2 from each of the terms in the numerator, while tossing a factor of 1 into the denominator.

$$D_n = \frac{1 \cdot 3 \cdot 5 \cdot \ldots \cdot (2n-5) \cdot 2^{n-2}}{1 \cdot 2 \cdot 3 \cdot 4 \cdot \ldots \cdot (n-1)}$$

Next he multiplied the numerator by the product $2 \cdot 4 \cdot 6 \cdot \ldots \cdot (2n-4)$, and the denominator by $1 \cdot 2 \cdot 3 \cdot \ldots \cdot (n-2) \cdot 2^{n-2}$, which is, of course, the same quantity.

$$D_n = \frac{1 \cdot 3 \cdot 5 \cdot \ldots \cdot (2n-5) \cdot 2^{n-2}}{1 \cdot 2 \cdot 3 \cdot 4 \cdot \ldots \cdot (n-1)} \cdot \frac{2 \cdot 4 \cdot 6 \cdot \ldots \cdot (2n-4)}{1 \cdot 2 \cdot 3 \cdot \ldots \cdot (n-2) \cdot 2^{n-2}}$$

The powers of two cancel, and by rearranging the terms in the numerator we can see that

$$D_n = \frac{(2n-4)!}{(n-1)!(n-2)!}$$

$$= \frac{1}{2n-3} \cdot \frac{(2n-3)!}{(n-1)!(n-2)!}$$

$$= \frac{1}{2n-3} \cdot \binom{2n-3}{n-1}.$$

Various other similar formulas are also possible, such as

$$D_n = \frac{1}{n-1} \cdot \binom{2n-4}{n-2}.$$

This technique of multiplying by a factorial and a power of two in order to "fill in the gaps" in a product of odd numbers is often useful in simplifying products, and is well worth remembering.

A summary of the problem-solving rules of thumb we have encountered so far:

[1] Start by working out the first few "small" cases, and look for a pattern. If you can guess the answer, you may be able to prove it by induction.

[2] If you can't spot the pattern, try for a recursion formula. That is, try to come up with a way of solving any given instance of the

problem by solving one or more smaller instances of the same problem.

[3] If you've got a recursion formula but aren't sure what to do with it, or if you're unable to find a recursion formula at all, try introducing a generating function the coefficients of which are the values you're interested in, and try to manipulate it to your advantage.

With that, we'll move on to the next chapter.

People who are interested in learning more about generating functions should read [Knuth], vol. 1, section 1.2.9. Knuth gives several additional rules for manipulating generating functions, and also discusses briefly the question of convergence. For example, $1+x+x^2+ \cdots = 1/(1-x)$ if and only if $|x|<1$. Two remarks from Knuth are particularly worth noting: " . . . it often does not pay to worry about convergence of the series when we work with generating functions, since we are only exploring possible approaches to the solution of some problem. When we discover the solution by *any* means, however sloppy they might be, it may be possible to justify the solution independently" (for instance by mathematical induction). "Furthermore it can be shown that most (if not all) of the operations we do with generating functions can be rigorously justified without regard to the convergence of the series."

Suppose we have a set of N objects that have various properties α, β, γ, ..., λ. Each of the objects may have any or none of the properties. Let N_α be the number of objects that have property α. Some of these objects may have other properties in addition to property α; that doesn't matter. (In fact, that's the whole idea!) Similarly, let N_β be the number of objects that have property β, and so on. Let $N_{\alpha\beta}$ be the number of objects that have <u>both</u> property α <u>and</u> property β, $N_{\alpha\gamma}$ the number that have properties α and γ, etc. $N_{\alpha\beta\gamma...\lambda}$ is the number of objects with *all* the properties. Given all this information, we wish to find N_0, the number of objects that have *none* of the properties.

The general formula for computing this is called the <u>Principle of Inclusion and Exclusion</u> (or sometimes PIE for short), and is the following:

$$N_0 = N - N_\alpha - N_\beta - N_\gamma - \cdots - N_\lambda$$
$$+ N_{\alpha\beta} + N_{\alpha\gamma} + N_{\beta\gamma} + \cdots + N_{\kappa\lambda}$$
$$- N_{\alpha\beta\gamma} - N_{\alpha\beta\delta} - \cdots$$
$$\vdots$$
$$\pm N_{\alpha\beta\gamma...\lambda}$$

We will eventually prove this (in two different ways, no less!), but first let's take a look at some examples. After all, it looks as though we need to know a heck of a lot of information in order to compute N_0; wouldn't it be easier to compute it directly? As we'll see, it is sometimes much easier to compute the various N_α than it is to compute N_0.

January 24. As our first example, suppose you've written n letters, and have addressed n envelopes to go with them. At this point you leave the room, and someone who can't read (*e.g.*, a monkey) wanders in and proceeds to put the letters into the envelopes at random, one letter per envelope. In how many ways can this be done such that no letter is in the right envelope? Equivalently, we can take the numbers $1, 2, 3, \ldots, n$, and look at a permutation $i_1, i_2, i_3, \ldots, i_n$. How many such permutations have $i_k \neq k$ for all k?

G. Pólya et al., *Notes on Introductory Combinatorics*, Modern Birkhäuser Classics,
DOI 10.1007/978-0-8176-4953-1_4, © Birkhäuser Boston, a part of Springer
Science+Business Media, LLC 2010

To solve this using PIE, we let α be the property that $i_1 = 1$ (*i.e.*, the first letter is in the first envelope), β the property that $i_2 = 2, \ldots$, and λ the property that $i_n = n$. In the PIE formula, N is the total number of permutations, which we know is $n!$. What about N_α? It is the number of permutations with $i_1 = 1$. Since i_1 must be 1 and the remaining $n-1$ elements can be in any order, we are counting all permutations of those $n-1$ elements, and there are $(n-1)!$ such permutations. Similarly, $N_\beta = N_\gamma = \cdots = N_\lambda = (n-1)!$. By the same reasoning $N_{\alpha\beta}$ counts permutations in which $i_1 = 1$ and $i_2 = 2$, with the remaining $n-2$ elements in any order, so $N_{\alpha\beta} = (n-2)!$, and likewise $N_{\alpha\gamma} = N_{\beta\gamma} = \cdots = N_{\kappa\lambda} = (n-2)!$. Continuing in this way, we determine that $N_{\alpha\beta\gamma} = (n-3)!$, $N_{\alpha\beta\gamma\delta} = (n-4)!, \ldots$, and finally $N_{\alpha\beta\gamma\ldots\lambda} = (n-n)! = 0! = 1$.

So far so good, and these numbers are certainly a lot easier to compute than N_0. Now, how many terms do we have that are equal to $(n-2)!$? Well, each pair of properties ζ and η contribute a term $N_{\zeta\eta} = (n-2)!$; how many such pairs are there? Since there are n properties, there are $\binom{n}{2}$ pairs of properties. Similarly, there are $\binom{n}{3}$ subsets of three properties, each of which contributes a term of $(n-3)!$, and so forth. The PIE formula gives us

$$N_0 = n! - \binom{n}{1}(n-1)! + \binom{n}{2}(n-2)! - \binom{n}{3}(n-3)! + \cdots + (-1)^n\binom{n}{n}0!$$

$$= n! - \frac{n!}{1!(n-1)!}(n-1)! + \frac{n!}{2!(n-2)!}(n-2)!$$

$$- \frac{n!}{3!(n-3)!}(n-3)! + \cdots + (-1)^n\frac{n!}{n!0!}0!$$

$$= n! \cdot (1 - \frac{1}{1!} + \frac{1}{2!} - \frac{1}{3!} + \cdots + (-1)^n\frac{1}{n!}).$$

This should look familiar from calculus, whence we know that

$$e^x = 1 + \frac{x}{1!} + \frac{x^2}{2!} + \frac{x^3}{3!} + \cdots.$$

Our series isn't infinite, but for n large, $N_0 \approx n!e^{-1}$. The probability that none of the letters is in its corresponding envelope is the number of such configurations divided by the number of possible configurations ($n!$), so this probability is approximately e^{-1}, which is $0.36787944 \ldots$ (Actually, n doesn't have to be very large for this to be quite accurate. When $n = 9$ it is already accurate to six decimal places. This means that the probability of getting ten letters each in a wrong envelope is not significantly different from that for a thousand letters.)

Let's try a more complicated example. Suppose you want to know how many prime numbers are less than 1,000,000. Nowadays you could grind it out on a computer by factoring every number from 1 to 1,000,000, but there's a method that requires much less computation. Let's assume you know (or can compute) the prime numbers up to 1000. Call them $p_1, p_2, p_3, \ldots, p_l$. Let α be the property of being evenly divisible by p_1, β the property of being divisible by $p_2, \ldots$, and λ the property of being divisible by p_l. Our set of size N will be the set of integers from 1 to N, with $N = 1,000,000$. In the general case, p_l would be the largest prime less than or equal to $\sqrt{N}$. We'll see later why this is so.

What is N_α? That is to say, how many numbers from 1 to N are divisible by p_1? Let kp_1 be the largest such number. Then $(k+1)p_1$ must be strictly greater than N, and we have

$$p_1, 2p_1, 3p_1, \ldots, kp_1 \leq N < (k+1)p_1.$$

So $N_\alpha = k$, where k is the unique integer such that

$$k \leq \frac{N}{p_1} < k+1.$$

The notation for this, as was mentioned on page 21, is $k = \lfloor N/p_1 \rfloor$, where $\lfloor x \rfloor$ denotes the largest integer less than or equal to x. So $N_\alpha = \lfloor N/p_1 \rfloor$, $N_\beta = \lfloor N/p_2 \rfloor, \ldots$, and $N_\lambda = \lfloor N/p_l \rfloor$.

Next we need to realise a fundamental property of prime numbers. If a number j is divisible by each of two distinct primes p_a and p_b, then j is divisible by the product $p_a p_b$. [Points to ponder: Why is this so? Why is it not necessarily true if p_a and p_b are not both prime? Can you find pairs of composite (non-prime) numbers for which it is true? What if $p_a = p_b$?] So $N_{\alpha\beta} = \lfloor N/p_1 p_2 \rfloor$, and so on.

We now know enough to be able to compute N_0, but just what does that give us? Well, none of the prime numbers $> \sqrt{N}$ and $\leq N$ will be divisible by any primes $\leq \sqrt{N}$, so they will be counted in N_0. N_0 will also count the number 1, since it is not divisible by any prime. It will *not* count the primes $p_1, p_2, \ldots, p_l$, since each is divisible by itself and has thus been excluded. Most important, N_0 will not count any composite numbers. To see that this is so, consider any composite number n, and consider its <u>smallest</u> factor, p. We know p must be a prime, because otherwise it would have a factor which would be a smaller factor of n. So $n = p\bar{p}$, where $\bar{p}$ may or may not be prime. Since p is the smallest factor of n, we know that $\bar{p} \geq p$. Therefore $n \geq p^2$, and thus $p \leq \sqrt{n}$. Hence any composite number $n \leq N$ must have at least one prime factor p such that $p \leq \sqrt{n} \leq \sqrt{N}$. This prime p, being less than $\sqrt{N}$, must be one of the primes $p_1, p_2, p_3, \ldots, p_l$. Since all multiples of these primes have been excluded from N_0, the composite number n will not be counted. And what *is* N_0? By the PIE,

$$
\begin{aligned}
N_0 = N &- \lfloor N/p_1 \rfloor - \lfloor N/p_2 \rfloor - \cdots - \lfloor N/p_l \rfloor \\
&+ \lfloor N/p_1 p_2 \rfloor + \lfloor N/p_1 p_3 \rfloor + \cdots \\
&- \lfloor N/p_1 p_2 p_3 \rfloor - \cdots \\
&\vdots \\
&+ (-1)^l \lfloor N/p_1 p_2 p_3 \ldots p_l \rfloor.
\end{aligned}
$$

The first problem on the fourth homework assignment was to find the number of primes between 10 and 100, first without using the above formula, then using it. Without it, it's fairly simple to tabulate the primes in an easy-to-read format:

11		31	41		61	71		
13	23		43	53		73	83	
17		37	47		67			97
19	29			59		79	89	

and we see there are 21 such primes. Using the formula, we have $l = 4$ and $p_1 = 2$, $p_2 = 3$, $p_3 = 5$, and $p_4 = 7$. Plugging these values in, we get

$$N_0 = 100 - \lfloor 100/2 \rfloor - \lfloor 100/3 \rfloor - \lfloor 100/5 \rfloor - \lfloor 100/7 \rfloor$$
$$+ \lfloor 100/6 \rfloor + \lfloor 100/10 \rfloor + \lfloor 100/14 \rfloor$$
$$+ \lfloor 100/15 \rfloor + \lfloor 100/21 \rfloor + \lfloor 100/35 \rfloor$$
$$- \lfloor 100/30 \rfloor - \lfloor 100/42 \rfloor - \lfloor 100/70 \rfloor - \lfloor 100/105 \rfloor$$
$$+ \lfloor 100/210 \rfloor$$

$$= 100 - 50 - 33 - 20 - 14$$
$$+ 16 + 10 + 7 + 6 + 4 + 2$$
$$- 3 - 2 - 1 - 0$$
$$+ 0$$

$$= 22.$$

We subtract 1 to account for the number 1 being included in N_0 and conclude that there are 21 primes between 10 and 100. The answers match. It may seem that evaluating the PIE formula was more work than creating the table, but that's just because of the size of the example. Try it with N = 1000 and see which way you think is easier! Note also that, as N increases, more and more terms in the PIE formula will be zero, because the products of primes in the denominators will exceed N.

An important variation on the preceding result can be found by taking N = $n \geq 2$ and letting $p_1, p_2, p_3, \ldots, p_l$ be, not *all* primes $\leq \sqrt{n}$, but just the distinct prime factors of n itself (some of which may be $> \sqrt{n}$). Then N/p_1, N/p_2, $N/p_1 p_2$, etc. are all integers, which means we can drop the "$\lfloor \ \rfloor$" symbols and get

$$N_0 = n - n/p_1 - n/p_2 - \cdots - n/p_l$$
$$+ n/p_1 p_2 + n/p_1 p_3 + \cdots$$
$$- n/p_1 p_2 p_3 - \cdots$$
$$\vdots$$
$$+ (-1)^l n/p_1 p_2 p_3 \ldots p_l$$

$$= n(1 - \frac{1}{p_1})(1 - \frac{1}{p_2})(1 - \frac{1}{p_3}) \ldots (1 - \frac{1}{p_l}).$$

This is the number of integers from 1 to n that are <u>relatively prime</u> to n, *i.e.*, have no factors in common with n. This is sometimes referred to as the <u>Euler-totient</u> function, or simply the totient; Euler's notation for it was $\varphi(n)$. For example, the prime factors of 36 are 2

and 3, so

$$\varphi(36) = 36(\tfrac{1}{2})(\tfrac{2}{3}) = 12,$$

which tells us that there are 12 numbers between 1 and 36 that are relatively prime to 36. We can check this: the numbers are 1, 5, 7, 11, 13, 17, 19, 23, 25, 29, 31, and 35.

Now that we've seen how PIE can be useful, how about a proof? We promised to present two proofs eventually; here's the first. Consider an object that has k of the properties. N counts it once. The sum $N_\alpha + N_\beta + N_\gamma + \cdots + N_\lambda$ counts it exactly k times. $N_{\alpha\beta} + N_{\alpha\gamma} + \cdots + N_{\kappa\lambda}$ counts it once for each way of selecting two properties from among the k, which is $\binom{k}{2}$ times. $N_{\alpha\beta\gamma} + N_{\alpha\beta\delta} + \cdots$ counts it once for each way of selecting three of the k properties, which is $\binom{k}{3}$ times, and so on. Altogether the PIE formula will count it $1 - k + \binom{k}{2} - \binom{k}{3} + \cdots \pm \binom{k}{n}$ times. But by the binomial theorem this is simply $(1-1)^k$, which is zero if $k \geq 1$. So any object with one or more properties is counted exactly zero times in N_0, which is what we want. What if $k = 0$, i.e., the object has none of the properties? In that case, N counts it once, and none of the other terms counts it at all, so N_0 counts it exactly once. That completes our proof.

The preceding proof is valid, but it's not aesthetically pleasing. The Principle of Inclusion and Exclusion is an extremely important result in set theory; surely it deserves to be proved using set theory instead of combinatorics! Very well, but first we must define a bit of formal logic.

January 26. Formal logic dates back to Aristotle, and is based on syllogisms. For instance, if we accept the premises "if A then B" and "if B then C", where A, B, and C represent assertions, then it follows that "if A then C". This line of reasoning is a syllogism. We can see it pictorially. First we look at the set of objects for which A is true.

Now, "if A then B" means that anything in the set A must also be in the set B, as shown on the following page.

Similarly, anything in the set B must also be in the set C:

and it is immediately obvious that everything in set A is in set C. (As Pólya put it, "If you listen to the words, you will probably admit it, but if you look at the figures it becomes completely evident.")

Next we consider the notion of a function. For example, $f(x) = x^2-1$ is a function of x. Here x is required to be a number; this need not always be the case. For instance, we could have let $f(x)$ = age of x. Here x is anything that has an age, such as people, trees, stars, etc. Now we shall apply some of this to set theory. We'll use U to represent the <u>universe of discourse,</u> i.e., the totality of things in which we are interested. A, B, C, etc. will represent subsets of U, and x, y, z, etc. will be individual objects in U. For any set A, we define the <u>characteristic function</u> attached to A, written A(x), as follows:

$$A(x) = \begin{cases} 1 & \text{if } x \text{ belongs to A;} \\ 0 & \text{if } x \text{ does not belong to A.} \end{cases}$$

For example, let U be the set {1,2,3,4,5,6} and let A be the set of divisors of 6. Then A(1)=1, A(2)=1, A(3)=1, A(4)=0, A(5)=0, and A(6)=1. Notice that U(x) = 1 for all x. The null (empty) set ϕ has the characteristic function $\phi(x)$ = 0 for all x.

The <u>complement</u> of A, which we'll denote by $\overline{A}$, consists of exactly those objects that are *not* in A. It thus has the characteristic function $\overline{A}(x) = 1 - A(x)$.

The <u>intersection</u> of two sets A and B, written A∩B, consists of those objects that are included in *both* A *and* B. If C = A∩B, then $C(x) = A(x)B(x)$.

The <u>union</u> of A and B, written A∪B, consists of those objects that are in *either* A *or* B (including objects that are in both). If D = A∪B, then the only objects *not* in D are those neither in A nor in B; *i.e.*, the complement of D is the intersection of the complements of A and B. We get therefore that the characteristic function of D is $D(x) = 1 - (1-A(x))(1-B(x)) = A(x) + B(x) - A(x)B(x)$.

Finally, the number of elements in a set A is the summation of the characteristic function over all elements in U:

$$\sum_{x \text{ in } U} A(x).$$

Note: This operation works only if it is a <u>finite</u> sum, so U must be a finite set.

So we can now transform operations on sets into arithmetic operations. Let's apply this to the PIE. We'll let A be the set of objects with property α, B the set of objects with property β, ..., and L the set of objects with property λ. We let $\hat{N}$ be the set of objects having none of the properties $\alpha, \beta, \ldots, \lambda$. What is the characteristic function of $\hat{N}$? Every element of $\hat{N}$ is contained in the complement of each of the sets A, B, ..., L. That is, $\hat{N}$ is the intersection of the complements of all of the individual sets. [Note: To simplify the equations that follow, we'll use A as an abbreviation of $A(x)$, except where it might be confused with the set A.] So the characteristic function is

$$\hat{N} = (1-A)(1-B)(1-C) \ldots (1-L).$$

We want to know how many elements are in the set $\hat{N}$, since that is N_0, the number of elements having none of the properties. So we take the summation of the characteristic function and find that this number is

$$\sum_{x \text{ in } U} (1-A)(1-B)(1-C)\ldots(1-L)$$

$$= \sum_{x \text{ in } U} (1 - A - B - C - \cdots - L$$
$$+ AB + AC + \cdots + KL$$
$$- ABC - \cdots$$
$$\vdots$$
$$\pm ABC\ldots L\,)$$

$$= \sum_{x \text{ in } U} 1 - \sum_{x \text{ in } U} A - \sum_{x \text{ in } U} B - \cdots$$
$$+ \sum_{x \text{ in } U} AB + \cdots$$
$$\vdots$$
$$\pm \sum_{x \text{ in } U} ABC\ldots L.$$

Since $\sum_{x \text{ in } U} AB$, for example, is the size of the set $A \cap B$, which in turn is the set of objects with both properties α and β, this is the same as $N_{\alpha\beta}$ in our earlier notation. So we conclude that

$$\sum_{x \text{ in } U} \hat{N}(x) = N_0 = N - N_\alpha - N_\beta - N_\gamma - \cdots - N_\lambda$$
$$+ N_{\alpha\beta} + N_{\alpha\gamma} + N_{\beta\gamma} + \cdots + N_{\kappa\lambda}$$
$$- N_{\alpha\beta\gamma} - N_{\alpha\beta\delta} - \cdots$$
$$\vdots$$
$$\pm N_{\alpha\beta\gamma\ldots\lambda}.$$

That's it for PIE. We'll be seeing more of it later on.

<table>
<tr><td>5</td><td>Stirling Numbers</td></tr>
</table>

5   Stirling Numbers

We now come to a somewhat esoteric set of numbers called <u>Stirling numbers</u> (after mathematician James Stirling). There are two kinds of Stirling numbers; they are called, appropriately enough, <u>Stirling numbers of the first kind</u> and <u>Stirling numbers of the second kind</u>. We'll start with the second kind.

We define S_k^n to be the number of ways to divide a set of size n into k <u>non-overlapping</u>, <u>non-empty</u> subsets whose union is the whole set. Such a division is called a <u>partition</u> into k subsets.

Incidentally, you should be warned that there is no "standard" notation for Stirling numbers. The notation we'll be using is one of the more common ones, but if you read any texts on this subject you should be prepared to see various other notations. Among the more common notations are S_n^k (confusing!) and $\{_k^n\}$.

Let's look at a few sample values in order to get a feel for these numbers. We start by observing that $S_k^n = 0$ unless $1 \leq k \leq n$. (Obviously, we cannot partition n objects into fewer than 1 set, nor can we form more than n subsets and still have all the sets be non-empty.) We also observe that, if $k = 1$, there's only one way to "divide" the set. The same is true if $k = n$; each object must belong to a different subset, and the order of the subsets is not being considered. So the first complicated value is S_2^3. The set $\{a,b,c\}$ can be divided into $\{a,b\}$ and $\{c\}$, or $\{a,c\}$ and $\{b\}$, or $\{b,c\}$ and $\{a\}$. So $S_2^3 = 3$.

What about S_2^4? If we are going to partition a set of 4 objects into 2 subsets, the subsets must be either of sizes 3 and 1 or of sizes 2 and 2. In the first case, there are 4 choices for the element in the set of size 1. The second case is a bit tricky—we're tempted to say that there are $\binom{4}{2} = 6$ choices for the pair of objects to be placed in the first subset, but this would be incorrect. Choosing $\{a,b\}$ for the first subset is equivalent to choosing $\{c,d\}$, since the order of the two subsets is irrelevant. The correct way to count these partitions is to look at some element, say a. It must be in one or the other subset, and it doesn't matter which since the two subsets are symmetrically equivalent. Whichever it's in, there are 3 choices for the other

G. Pólya et al., *Notes on Introductory Combinatorics*, Modern Birkhäuser Classics,
DOI 10.1007/978-0-8176-4953-1_5, © Birkhäuser Boston, a part of Springer
Science+Business Media, LLC 2010

element of that subset. Having made that choice, we've completely determined the partition, since the other two elements must go in the other subset. So there are three partitions of 4 objects into two subsets of size 2. Altogether then, $S_2^4 = 7$.

For S_3^4, the subsets must be of sizes 2, 1, and 1. There are $\binom{4}{2} = 6$ ways to choose the subset of size 2, and each such choice completely determines the partition. (Note that, unlike the situation encountered in computing S_2^4, all 6 cases are now different.) So $S_3^4 = 6$.

Pólya drew up a table showing these values, and assigned as homework the calculation of the 5th and 6th rows of the table. (The values were then to be checked using the recursion formula which we'll get to next.) Here is the table, extended through row 10.

S_k^n										
n \ k	1	2	3	4	5	6	7	8	9	10
1	1	0	0	0	0	0	0	0	0	0
2	1	1	0	0	0	0	0	0	0	0
3	1	3	1	0	0	0	0	0	0	0
4	1	7	6	1	0	0	0	0	0	0
5	1	15	25	10	1	0	0	0	0	0
6	1	31	90	65	15	1	0	0	0	0
7	1	63	301	350	140	21	1	0	0	0
8	1	127	966	1701	1050	266	28	1	0	0
9	1	255	3025	7770	6951	2646	462	36	1	0
10	1	511	9330	34105	42525	22827	5880	750	45	1

A typical line of reasoning for computing one of the above values by inspection (as opposed to by recursion) would be the following one for S_3^5. The three subsets must be of sizes 3, 1, and 1, or of sizes 2, 2, and 1. In the first case, we have $\binom{5}{3} = 10$ choices for the set of size 3, which completely determines the partition. In the second case, we have 5 choices for the element in the set of size 1, and the remaining 4 elements must be partitioned into two sets of size 2, which we know (from having computed S_2^4) can be done in any of 3 ways. So there are $3 \cdot 5 = 15$ ways to partition 5 objects into sets of sizes 2, 2, and 1; together with the other 10 we have 25 partitions.

Generating this table by inspection would become tedious after about the 6th or 7th row, so let's determine a recursion formula for these numbers. Consider the transition from n to $n+1$. For example, suppose you step into a room with n other people. All $n+1$ of you are to be separated into k non-empty groups. By definition there are S_k^{n+1} ways this can be accomplished. By approaching the matter from a different direction, however, we can get another formula, which must therefore be equal to S_k^{n+1}. There are two possibilities. First, you could be antisocial and form a group all by yourself. The other n people would then have to form $k-1$ groups. There are S_{k-1}^n ways for them to do this. Alternatively, you could decide you feel like having company. In this case the other n people would form k groups, and you would then join one of their groups. There are k choices for which group you decide to join, and S_k^n ways for the other people to have formed the k groups, for a total of kS_k^n possibilities. Adding the antisocial case, we find

$$S_k^{n+1} = S_{k-1}^n + kS_k^n.$$

This formula can be checked using the first few rows of the table, which we have already computed by inspection. (The remaining rows were, of course, computed using the recursion formula.)

A digression: If you want a bit of practice with mathematical induction and/or this recursion formula, try using them to prove the following pair of hypotheses for $n \geq 1$.

$$S_2^n = 2^{n-1} - 1 \quad \text{and} \quad S_{n-1}^n = \binom{n}{2}$$

January 31. As an example, suppose we have k different colors of paint, and we wish to paint n houses. (Each house is to be painted a single color.) In how many ways can we do this?

The first house can be painted in any of k different ways. Independent of the color we choose for the first house, the second house can also be painted in any of k different ways, and so on for each house. So there are $k \cdot k \cdot k \cdot \ldots \cdot k = k^n$ ways. But this includes cases where not all of the k colors are used; *e.g.*, all the houses might be painted blue. How many ways actually use all k colors?

This looks like a job for PIE. Let α be the property that no house is painted with the 1st color, β the property that no house is painted with the 2nd color, ..., and λ the property that no house is painted with the kth color. We want the number of ways to paint the houses such that all the colors are used, *i.e.*, color assignments that have none of the properties $\alpha, \beta, \ldots, \lambda$. We recall from the previous chapter that

$$N_0 = N - N_\alpha - N_\beta - N_\gamma - \cdots - N_\lambda$$
$$+ N_{\alpha\beta} + N_{\alpha\gamma} + N_{\beta\gamma} + \cdots + N_{\kappa\lambda}$$
$$- N_{\alpha\beta\gamma} - N_{\alpha\beta\delta} - \cdots$$
$$\vdots$$
$$\pm N_{\alpha\beta\gamma\ldots\lambda}.$$

N we know is k^n. N_α is the number of ways to paint the houses without using color #1. Since each house can be any of $(k-1)$ colors, $N_\alpha = (k-1)^n$. The same goes for N_β, N_γ, etc., for a total of k such terms. Similarly, $N_{\alpha\beta} = (k-2)^n$, since each house can be any of the $(k-2)$ remaining colors. There are $\binom{k}{2}$ such terms. Carrying on in this fashion, we find

$$N_0 = k^n - \binom{k}{1}\cdot(k-1)^n + \binom{k}{2}\cdot(k-2)^n - \binom{k}{3}\cdot(k-3)^n + \cdots + (-1)^k\binom{k}{k}\cdot0^n.$$

As usual, we'll check this formula on a few special cases, just to see whether we've made any obvious mistakes. For instance, if $k = 1$ and $n \geq 1$, there should be only one way to paint the houses using the single available color. The formula yields

$$N_0 = 1^n - \binom{1}{1}\cdot0^n$$

$$= 1,$$

which checks. How about $n = 1$ and $k \geq 2$? There's no way to paint one house so as to use two or more colors (since we're restricting ourselves to a single color per house), so the formula should yield zero.

$$N_0 = k - \binom{k}{1} \cdot (k-1) + \binom{k}{2} \cdot (k-2) - \binom{k}{3} \cdot (k-3) + \cdots + (-1)^k \binom{k}{k} \cdot 0$$

$$= \sum_{s=0}^{k-1} (-1)^s \binom{k}{s} \cdot (k-s) \qquad \text{[the last term in the previous line } = 0 \text{ and is omitted from the summation]}$$

$$= \sum_{s=0}^{k-1} \frac{k! \, (k-s)}{s! \, (k-s)!} (-1)^s$$

$$= \sum_{s=0}^{k-1} \frac{k \cdot (k-1)!}{s! \, (k-s-1)!} (-1)^s$$

$$= \sum_{s=0}^{k-1} k \cdot \binom{k-1}{s} \cdot (-1)^s$$

$$= k \sum_{s=0}^{k-1} \binom{k-1}{s} \cdot (-1)^s$$

$$= k \cdot (1-1)^{k-1}$$

$$= k \cdot 0^{k-1}$$

So if $k \geq 2$, this indeed comes out zero. Note that, if $k = 1$, we once again find ourselves relying on 0^0 being equal to 1. As a final special case, consider $n = 0$. Since $0^0 = 1$, the formula turns into

$$N_0 = 1 - \binom{k}{1} + \binom{k}{2} - \binom{k}{3} + \cdots + (-1)^k \binom{k}{k}$$

$$= (1-1)^k,$$

which equals zero if $k \geq 1$. Since there is no way to paint zero houses so as to use one or more colors, this checks. If $k = 0$, we find there is exactly one way to paint zero houses using zero colors, which sounds reasonable, too.

Okay, so the formula looks good. Let's approach the problem in a different way and see if we can learn anything. When we paint the n houses, we could do it by first partitioning the houses into k sets—the set of houses painted the 1st color, the set painted the 2nd color, etc. We require that each of these sets be non-empty, since each color is to be used on at least one house. Therefore there are S_k^n ways to partition the houses. Having done so, we then have $k!$ ways to assign the k colors to the k sets, so there are $S_k^n \cdot k!$ ways to paint the houses using all k colors. But this means that

$$S_k^n \, k! = k^n - \binom{k}{1}(k-1)^n + \binom{k}{2} \cdot (k-2)^n - \binom{k}{3}(k-3)^n + \cdots + (-1)^k \binom{k}{k} \cdot 0^n.$$

(Checking this formula when n is equal to 5 and 6 was assigned as homework. It is a fairly straightforward procedure given the table for S_k^n which we produced earlier, so we won't go into it here.)

For our next bit of fun, let's try to turn this into a generating function. It turns out that a generating function in which S_k^n is the coefficient of z^n is awkward. This very often happens when the coefficients increase particularly rapidly; though we don't normally need to worry about whether a generating function converges, it helps if it converges for at least *some* non-zero values of z. If the coefficients of the generating function are growing faster than the powers of z can shrink (for $z<1$), then it will not converge (*i.e.*, the sum will be infinite). When this sort of problem arises, it often pays to divide the coefficients by something that itself grows very quickly, namely $n!$. So, given a particular value of k, we let n vary and take the summation

$$\sum_{n=k}^{\infty} \frac{S_k^n}{n!} z^n = \frac{1}{k!} \cdot \sum_{n=k}^{\infty} \left[\frac{(kz)^n}{n!} - \binom{k}{1} \frac{((k-1)z)^n}{n!} + \binom{k}{2} \frac{((k-2)z)^n}{n!} - \cdots \right].$$

Since $S_k^n = 0$ for $0 \le n < k$ and since our previous formula holds there also, we can extend the summation to start at zero instead of k.

$$\sum_{n=0}^{\infty} \frac{S_k^n}{n!} z^n = \frac{1}{k!} \left[\sum_{n=0}^{\infty} \frac{(kz)^n}{n!} - \sum_{n=0}^{\infty} \binom{k}{1} \frac{((k-1)z)^n}{n!} + \sum_{n=0}^{\infty} \binom{k}{2} \frac{((k-2)z)^n}{n!} - \cdots \right]$$

Remembering that

$$\sum_{n=0}^{\infty} \frac{x^n}{n!} = 1 + \frac{x}{1!} + \frac{x^2}{2!} + \frac{x^3}{3!} + \cdots = e^x,$$

we get

$$\sum_{n=0}^{\infty} \frac{S_k^n}{n!} z^n = \frac{1}{k!} \left[e^{kz} - \binom{k}{1} e^{(k-1)z} + \binom{k}{2} e^{(k-2)z} - \cdots + (-1)^k \binom{k}{k} e^0 \right]$$

$$= \frac{(e^z - 1)^k}{k!}.$$

That's enough for now with regard to Stirling numbers of the second kind. Let's take a look at the first kind. Those of the second kind were defined using partitions of a set; those of the first kind are defined using cycles of a permutation. So it's time to define a few more terms.

Consider any permutation of n objects. We've been thinking of it as an ordering of the objects, one of $n!$ possible orderings. We can instead represent it by writing the numbers 1 through n to denote the n objects, and writing below k the number of the kth element of the ordering. For instance, if the first element of the set is placed fifth in the particular permutation, then we write 1 below 5. The general notation for this will be

$$\begin{pmatrix} 1 & 2 & 3 & 4 & \dots & n \\ i_1 & i_2 & i_3 & i_4 & \dots & i_n \end{pmatrix}.$$

For example, for $n = 6$, one possible permutation is

$$\begin{pmatrix} 1 & 2 & 3 & 4 & 5 & 6 \\ 3 & 5 & 6 & 4 & 2 & 1 \end{pmatrix}.$$

Here's the important concept: We can think of this as a *function* mapping the set $\{1,2,3,4,5,6\}$ onto itself. For this particular permutation, $f(1) = 3$, $f(2) = 5, \dots,$ and $f(6) = 1$. This functional interpretation of permutations will become more significant in the next chapter. For now, we're only interested in the cycles of the function. These are easier to understand if we look at the function graphically. Extending our earlier example, we can represent the function f by using arrows to indicate the operation; *e.g.*, $1 \rightarrow 3$ indicates $f(1) = 3$.

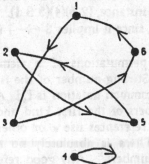

The graph becomes somewhat easier to read if we separate it into its independent parts.

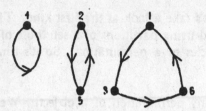

Each of these parts is called a <u>cycle</u>. A cycle of k elements (also called a "cycle of order k") represents a portion of the permutation that, if applied k times, restores the original ordering to those k elements. In our example, there is a cycle of order 3 containing the elements 1, 3, and 6. This tells us that $f(f(f(1))) = 1$, $f(f(f(3))) = 3$, and $f(f(f(6))) = 6$. Similarly, $f(f(2)) = 2$, $f(f(5)) = 5$, and $f(4) = 4$. Note the advantage of thinking of the permutation as a function, which permits us to perform it multiple times, an operation that might be hard to visualise in terms of reordering elements of a set. (As we mentioned before, we'll see more of this in the next chapter.)

A permutation can be completely specified by showing the cycles. The usual notation for this involves putting each cycle's elements inside a set of parentheses; our example would be written as

$$(1\ 3\ 6)(2\ 5)(4).$$

This represents the mapping

$$(1 \rightarrow 3, 3 \rightarrow 6, 6 \rightarrow 1)(2 \rightarrow 5, 5 \rightarrow 2)(4 \rightarrow 4)$$

Note that the cycle notation is not unique. The above permutation could be written as, for instance, $(2\ 5)(4)(3\ 6\ 1)$. But $(2\ 5)(4)(3\ 1\ 6)$ is a different permutation, since it implies $3 \rightarrow 1$ instead of $1 \rightarrow 3$.

The number of permutations of n elements that consist of precisely k cycles is the Stirling number of the first kind, which we'll write as $\mathcal{S}_k^n$. (Another common notation is $\left[\begin{smallmatrix} n \\ k \end{smallmatrix}\right]$. Also, some references use S for Stirling numbers of the first kind and $\mathcal{S}$ for those of the second kind, and other references use σ or other esoteric characters. As Knuth observes, "There is absolutely no agreement today on notation for Stirling's numbers." Any good reference will therefore go to great pains to describe the notation being used; look for this description any time you're reading up on Stirling numbers.)

To repeat then, we'll be using $\mathscr{S}_k^n$ for Stirling numbers of the first kind. Let's look at a few examples. For instance, when $n = 1$, there is only one permutation, namely $1 \rightarrow 1$, and it has a single cycle (of order one). Therefore $\mathscr{S}_1^1 = 1$. When $n = 2$, there are two permutations. One of them has $1 \rightarrow 1$ and $2 \rightarrow 2$, which in cycle notation is (1)(2). The other has $1 \rightarrow 2$ and $2 \rightarrow 1$, which in cycle notation is (1 2). So there is one permutation with one cycle and one with two cycles, and thus $\mathscr{S}_1^2 = \mathscr{S}_2^2 = 1$.

When $n = 3$ things begin to get more complicated. If $k = 1$, then all 3 elements have to be in a single cycle. There are only two possibilities, depending upon whether the cycle goes 'clockwise' or 'counterclockwise':

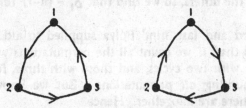

so $\mathscr{S}_1^3 = 2$. Meanwhile, if $k = 2$, the two cycles must consist of one cycle of order 1 and one of order 2, since together they must include all three elements. There are three choices for the element that is to be in the cycle of order 1; this choice having been made, there is only one way for the other two elements to go in the other cycle (remembering that (1)(2 3) and (1)(3 2) are the same permutation), so there are exactly 3 permutations of three elements with two cycles. Thus $\mathscr{S}_2^3 = 3$. Finally, if $k = 3$, each of the cycles must be of order 1. Hence each element must map to itself, and we get the "identity permutation" (1)(2)(3). Thus $\mathscr{S}_3^3 = 1$.

It was assigned as homework to calculate $\mathscr{S}_k^n$ for $n = 4$ and 5. As with S_k^n, the calculation was to be done first by inspection, and then using the recursion formula to be developed shortly. Pólya made things a bit easier by pointing out a few special cases. First, whenever $k = n$, each of the elements must be in a cycle by itself. There is only one such permutation—the identity permutation. So $\mathscr{S}_n^n = 1$ for all integers $n \geq 1$. Second, whenever $k = 1$, all n elements must be in a single cycle. We can "rotate" the cycle (or, if you prefer to think of it this way, we can write down the cycle and then turn the paper) to put some specific element, say n, at the top.

Having established this fixed point of view, we can then place the remaining $n-1$ elements in any order. Each such cycle will be different from the others, so we find that $\delta_1^n = (n-1)!$ for all $n \geq 1$.

The third and last 'hint' Pólya supplied to aid in doing the homework was that, if we count all the permutations with one cycle, and also those with two cycles, and those with three, four, five, etc., we end up counting *all* permutations. But we know how many permutations there are altogether. Hence

$$\sum_{k=1}^{n} \delta_k^n = n!$$

for all positive integers n.

As before, we won't go into all the details here regarding the computation by inspection of δ_k^n for $n = 4$ and 5. One typical value is δ_2^5, which we can compute as follows. The two cycles must either be one of order 3 and one of order 2, or else one of order 4 and one of order 1. In the latter case, there are 5 choices for the element in the order-1 cycle, and for each choice there are (by our earlier reasoning regarding δ_1^n) $(4-1)! = 6$ ways to arrange the order-4 cycle. In the other case, we have $\binom{5}{2} = 10$ choices for the order-2 cycle, and for each choice there are 2 ways to arrange the other elements in the order-3 cycle. So $\delta_2^5 = 5 \cdot 6 + 10 \cdot 2 = 50$. The following table shows δ_k^n for $n \leq 8$.

δ_k^n

n \ k	1	2	3	4	5	6	7	8
1	1	0	0	0	0	0	0	0
2	1	1	0	0	0	0	0	0
3	2	3	1	0	0	0	0	0
4	6	11	6	1	0	0	0	0
5	24	50	35	10	1	0	0	0
6	120	274	225	85	15	1	0	0
7	720	1764	1624	735	175	21	1	0
8	5040	13068	13132	6769	1960	322	28	1

Now let's try to find a recursion formula for these numbers. Suppose you enter a room and find n people already arranged in cycles. If you want to be alone, you can form a cycle by yourself. If there are to be k cycles altogether, then the other n people must be in $k-1$ cycles, and there are δ_{k-1}^n ways this can be done. But if you want to be sociable, you can join an existing cycle. In this case there must be k cycles already formed, and there are δ_k^n ways this can be done. For each such arrangement, how many ways are there for you to join one of the cycles? Well, for any particular cycle, you could step between any adjacent pair of people, so there are as many ways for you to join as there are people already in the cycle. So all told there are n ways for you to join the other n people. (Basically, you can step in front of any of the n people.) Adding the two cases, we get the recursion formula

$$\delta_k^{n+1} = \delta_{k-1}^n + n\delta_k^n.$$

Compare this to the recursion formula for Stirling numbers of the second kind:

$$S_k^{n+1} = S_{k-1}^n + kS_k^n.$$

Note the different coefficients on the second terms. You might also want to test out the formula for δ by using it to prove the special cases $\delta_1^n = (n-1)!$ and $\delta_{n-1}^n = \binom{n}{2}$.

February 2. We've looked at a generating function for Stirling numbers of the second kind. What about those of the first kind? For any particular value of n, we'll define

$$g_n(x) = \delta_1^n x + \delta_2^n x^2 + \delta_3^n x^3 + \cdots + \delta_n^n x^n.$$

(Note that we're taking n as fixed this time instead of k. We could also let k be fixed and sum over all values of n, getting a very different generating function, but we won't go into that in these notes.) We'll start by looking at the first few such functions and trying to spot a pattern.

$$g_1(x) = x$$
$$g_2(x) = x + x^2$$
$$g_3(x) = 2x + 3x^2 + x^3$$

Hmm, not much of a pattern there, or is there? Let's try factoring them.

$$g_1(x) = x \qquad\qquad = x$$
$$g_2(x) = x + x^2 \qquad\quad = x(1+x)$$
$$g_3(x) = 2x + 3x^2 + x^3 = x(1+x)(2+x)$$

Ah, that's much better! It looks as though

$$g_n(x) = x(x+1)(x+2)(x+3) \ldots (x+n-1).$$

Let's see if we can prove this conjecture. Some parts of it we can check at once. The coefficient of x^n will always be 1, which is indeed δ_n^n. The coefficient of x will be $1 \cdot 2 \cdot 3 \cdot \ldots \cdot (n-1) = (n-1)!$, which is δ_1^n. So it looks good; how do we prove it in general? Well, it's correct for $n = 1, 2,$ and 3, so let's try mathematical induction. We'll assume that $g_n(x) = x(x+1)(x+2) \ldots (x+n-1)$ for some value of n, and try to prove that $g_{n+1}(x) = x(x+1)(x+2) \ldots (x+n)$.

$$x(x+1)(x+2) \ldots (x+n) = [g_n(x)] \cdot (x+n)$$

$$= \delta_1^n x^2 + \delta_2^n x^3 + \cdots + \delta_{n-1}^n x^n + \delta_n^n x^{n+1}$$
$$\quad + n\delta_1^n x + n\delta_2^n x^2 + n\delta_3^n x^3 + \cdots + n\delta_n^n x^n$$

$$= n\delta_1^n x + (\delta_1^n + n\delta_2^n)x^2 + (\delta_2^n + n\delta_3^n)x^3 + \cdots + (\delta_{n-1}^n + n\delta_n^n)x^n + \delta_n^n x^{n+1}$$

Since $n\delta_1^n = n \cdot (n-1)! = n! = \delta_1^{n+1}$, and $\delta_n^n = 1 = \delta_{n+1}^{n+1}$, and (most importantly) $\delta_{k-1}^n + n\delta_k^n = \delta_k^{n+1}$ by the recursion formula, we find

$$x(x+1)(x+2)\ldots(x+n)$$

$$= \delta_1^{n+1} x + \delta_2^{n+1} x^2 + \delta_3^{n+1} x^3 + \cdots + \delta_n^{n+1} x^n + \delta_{n+1}^{n+1} x^{n+1}$$

$$= g_{n+1}(x).$$

This generating function is most often shown in a slightly different form, which we can derive by substituting $(-x)$ for x and multiplying both sides by $(-1)^n$:

$$(-1)^n \delta_1^n (-x) + (-1)^n \delta_2^n (-x)^2 + \cdots + (-1)^n \delta_n^n (-x)^n$$
$$= (-x)(-x+1)(-x+2)\ldots(-x+n-1)\cdot(-1)^n$$

$$\boxed{\delta_n^n x^n - \delta_{n-1}^n x^{n-1} + \delta_{n-2}^n x^{n-2} - \cdots + (-1)^{n-1} \delta_1^n x = x(x-1)(x-2)\ldots(x-n+1)}$$

Note that the right-hand side can be rewritten as $\binom{x}{n} n!$.

Let's go back for another look at S_k^n. Suppose we're painting n houses and have x colors available. In how many ways can we do this, *not necessarily* using all x colors? One way to look at it is to say that there are x choices for the color to use on the first house, and independent of that there are x choices for the color of the second, and so on for a total of x^n possibilities. Another way to look at it is to start by asking how many ways there are to paint the houses using *exactly one* color from among the x colors available. Obviously there are x ways. How many ways use exactly two colors? We first partition the houses into two subsets. We then choose any color (x possibilities) and paint all the houses in the first subset using that color. We then pick any *other* color ($x-1$ possibilities) and use it to paint the remaining houses. So there are $S_2^n x(x-1)$ ways to paint the houses using exactly two of the x colors. Similarly, we can deduce that there are $S_3^n x(x-1)(x-2)$ ways to paint them using exactly three colors, and so forth. We conclude that:

$$\boxed{x^n = S_1^n x + S_2^n x(x-1) + S_3^n x(x-1)(x-2) + \cdots + S_n^n x(x-1)(x-2)\ldots(x-n+1)}$$

Compare this formula with the previous one; the two are in some vague sense symmetric. The first one represents numbers of the form $\binom{x}{k}k!$ in terms of powers of x; the second represents powers of x in terms of numbers of the form $\binom{x}{k}k!$. It was in this form that Stirling originally developed these numbers. The expressions $\binom{x}{k}k!$ play a significant role in interpolation methods, which is one reason why Stirling numbers are considered worth investigating.

That's it for Stirling numbers, and chapter 5 of these notes. For more information on Stirling numbers, you may refer to [P-Sz] vol. 1, pp. 42-45 and pp. 224-229. (The former pages contain various exercises, while the latter discuss the solutions.)

6 | Pólya's Theory of Counting

February 2. Pólya's title for this part of the course was actually "Counting Configurations Non–Equivalent with Respect to a Given Permutation Group". Just about everybody else refers to it as "Pólya's Theory of Counting". This latter title is somewhat easier to remember, though not as indicative of the content. We'll stick to the simpler title; for the content, read on!

Since we're going to be dealing with permutation groups, we'll start by covering a bit of Group Theory. You'll recall from the previous chapter that we have ceased thinking of a permutation as merely an ordering of n objects. It can be more generally thought of as a function, a <u>transformation</u> of n elements. For instance, $\begin{pmatrix} 1 & 2 & 3 \\ 2 & 3 & 1 \end{pmatrix}$ is a function, call it $f(x)$, such that $f(1) = 2$, $f(2) = 3$, and $f(3) = 1$. A second permutation, say $\begin{pmatrix} 1 & 2 & 3 \\ 1 & 3 & 2 \end{pmatrix}$, represents another function, $g(x)$, such that $g(1) = 1$, $g(2) = 3$, and $g(3) = 2$. We shall use the notation

$$\begin{pmatrix} 1 & 2 & 3 \\ 1 & 3 & 2 \end{pmatrix}\begin{pmatrix} 1 & 2 & 3 \\ 2 & 3 & 1 \end{pmatrix}$$

to represent $g(f(x))$; that is, the permutation on the *right* is done *first*. Note this is different from $f(g(x))$, since $f(g(1)) = f(1) = 2$, whereas $g(f(1)) = g(2) = 3$. Meanwhile, $g(f(2)) = 2$ and $g(f(3)) = 1$, so

$$\begin{pmatrix} 1 & 2 & 3 \\ 1 & 3 & 2 \end{pmatrix}\begin{pmatrix} 1 & 2 & 3 \\ 2 & 3 & 1 \end{pmatrix} = \begin{pmatrix} 1 & 2 & 3 \\ 3 & 2 & 1 \end{pmatrix}. \qquad (*)$$

Recalling our cycle notation, this can be stated as

$$(1)(2\ 3)(1\ 2\ 3) = (2)(1\ 3).$$

Since the number of elements being permuted is small, we can look at these permutations geometrically and thereby get a better feel for what we're doing. Taking the vertices of an equilateral triangle as the elements being permuted, we find the permutation $(2)(1\ 3)$ corresponds to "flipping" the triangle around an axis going through vertex number 2 and through the center of the opposite edge:

G. Pólya et al., *Notes on Introductory Combinatorics*, Modern Birkhäuser Classics,
DOI 10.1007/978-0-8176-4953-1_6, © Birkhäuser Boston, a part of Springer
Science+Business Media, LLC 2010

$$\begin{pmatrix} 1 & 2 & 3 \\ 3 & 2 & 1 \end{pmatrix} \Leftrightarrow \qquad \rightarrow \qquad (i)$$

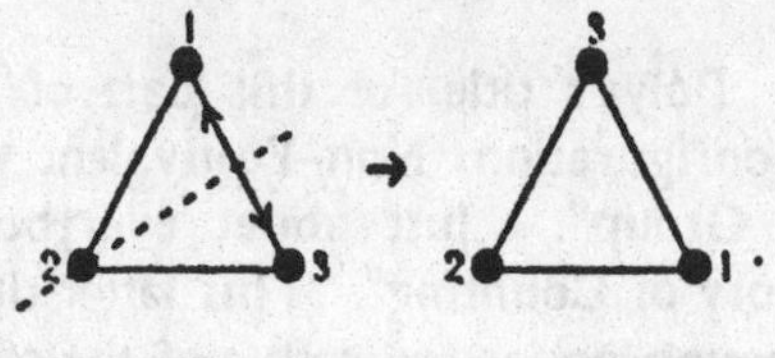

Similarly, we find that the other two permutations correspond to these rotations:

$$\begin{pmatrix} 1 & 2 & 3 \\ 2 & 3 & 1 \end{pmatrix} \Leftrightarrow \qquad \rightarrow \qquad (ii)$$

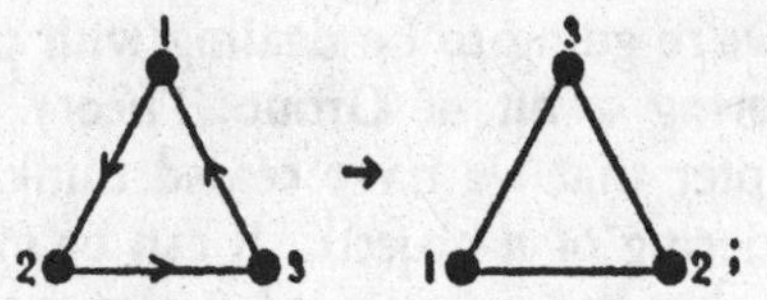

$$\begin{pmatrix} 1 & 2 & 3 \\ 1 & 3 & 2 \end{pmatrix} \Leftrightarrow \qquad \rightarrow \qquad (iii)$$

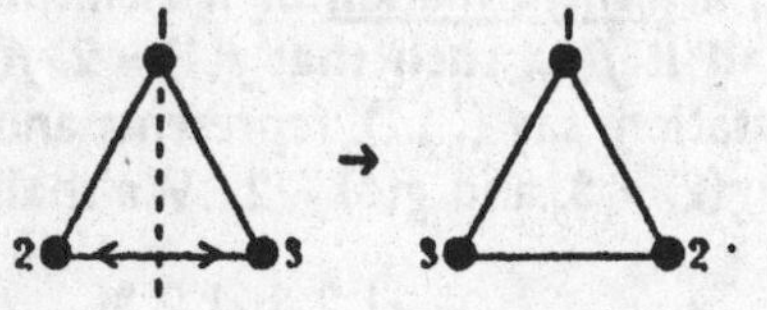

So the equation (∗) simply means that if we take an equilateral triangle and flip it as in (i), we get the same result as if we had rotated it as in (ii) and then flipped it as in (iii). The key fact is if we do any two transformations like this we end up with a situation that we could have achieved with a single motion.

A group is a set of operations (in this case, permutations) such that if you do one operation in the group, followed by another operation in the group (or possibly the same operation a second time), the combined operation is also in the group. In our example, having picked up the triangle, there are six ways for us to put it down. We have three choices for the vertex to be placed at the "top", and for each such choice we can place the triangle either "face-up" or "face-down". Thus there are six operations in this particular group. This is called the order of the group: the number of operations. The degree of the group is the number of objects involved; in our example the degree is three.

Of the six operations in this particular group, one involves "standing still":

$$\begin{pmatrix} 1 & 2 & 3 \\ 1 & 2 & 3 \end{pmatrix} \Leftrightarrow \qquad \rightarrow \qquad .$$

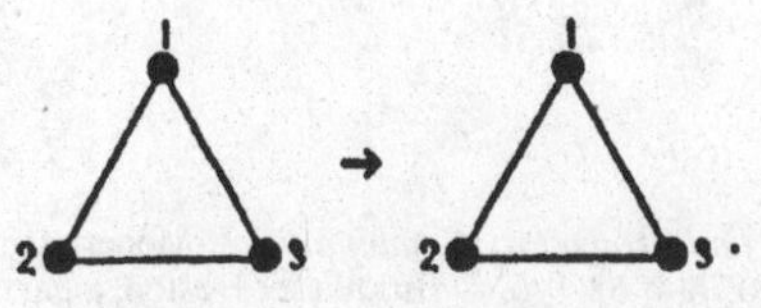

Three operations involve holding a vertex fixed and interchanging the other two (as in (*i*) and (*iii*) on the preceding page), and the remaining two operations involve rotation 120° in one direction or the other about the center (as in (*ii*)). So we have one case of the form (*a*)(*b*)(*c*) in cycle notation, three cases of the form (*a*)(*b c*), and two of the form (*a b c*). We shall represent a cycle of order *k* by the variable f_k. (Some texts use other notations.) A combination of cycles will be represented by a product; hence (1)(2 3) is denoted by $f_1 f_2$ and (1)(2)(3) by f_1^3. Finally, we add together the products corresponding to the various permutations, and then divide by the total number of permutations in the group. Since there is one permutation of the form f_1^3, three of the form $f_1 f_2$, and two of the form f_3, we come up with

$$\frac{f_1^3 + 3f_1 f_2 + 2f_3}{6}.$$

This is called the <u>cycle index</u> of the permutation group. We'll be defining it more precisely later on. It is perhaps the most important single concept in this course.

February 7. As a somewhat more interesting example, let's consider the rotations of a regular hexagon. A hexagon can be rotated in various ways such that it "coincides with itself". By "coincides" we mean that, if the hexagon is not labelled in some way, there is no way to distinguish the original position from the rotated position. Thus, for example, we do *not* include the rotation

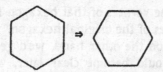

On the other hand, we *can* rotate the hexagon as indicated by the arrows in the leftmost figure below, or we can lift it out of the plane and rotate it 180° about either of the axes shown in dotted lines in the center and rightmost figures.

We wish to find the cycle index for this group. Well, to begin with, how many operations are in the group? That is, in how many different ways can we rotate a regular hexagon such that it coincides with itself? We pick some vertex and rotate the hexagon to place that vertex in the position of our choice; there are six possible positions. Having chosen where to place that vertex, we still have a decision left to make—we can place the hexagon "face up" or "face down". That is, we could flip the hexagon over or we could choose not to. Hence the total number of positions for the hexagon is 6×2 = 12. This is the *order* of the group of rotations of the hexagon.

Let us represent a rotation of the hexagon by noting which vertices take the places of which other vertices. For example, the rotation

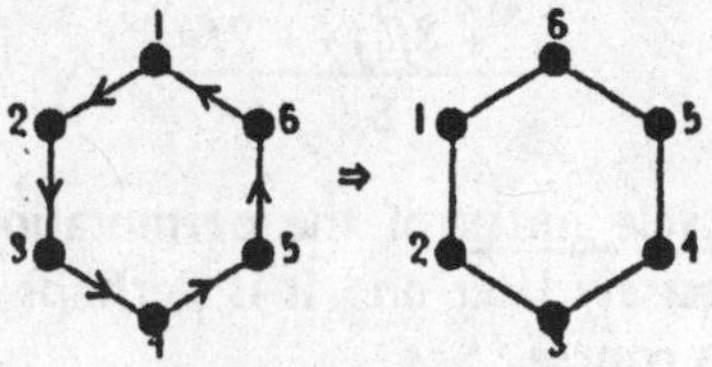

would be represented by the permutation

$$\begin{pmatrix} 1 & 2 & 3 & 4 & 5 & 6 \\ 2 & 3 & 4 & 5 & 6 & 1 \end{pmatrix}.$$

We could use other features of the hexagon instead of the vertices. If we used the midpoints of the edges, we'd get the same cycle index, since these midpoints form the vertices of another regular hexagon, so we are permuting the vertices of that hexagon in the same way we are permuting the edges of the original hexagon. If we looked at the three main diagonals, on the other hand, we'd get a different result. For reasons which should become clear later, we'll stick with the vertices. There are six vertices, so the *degree* of this permutation group is 6.

Let us now try to find all the permutations corresponding to rotations of the hexagon. One such, which we must be careful not to overlook, is the "rotation" consisting of no motion at all—the identity permutation.

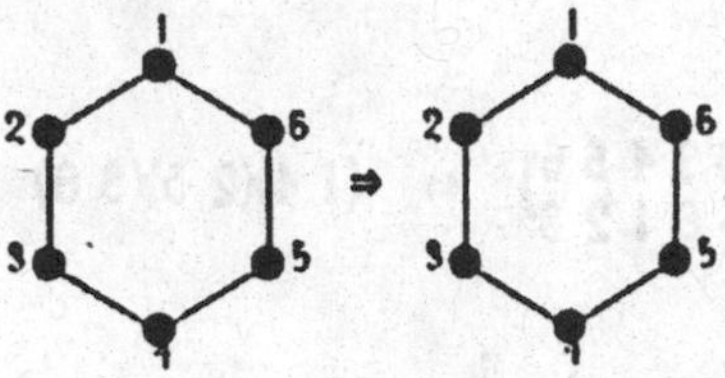

$$\Rightarrow \qquad \leftrightarrow \quad \begin{pmatrix} 1\ 2\ 3\ 4\ 5\ 6 \\ 1\ 2\ 3\ 4\ 5\ 6 \end{pmatrix} \quad \leftrightarrow \quad (1)(2)(3)(4)(5)(6)$$

As shown by the cycle notation at the far right, this permutation has six cycles of order 1, so it contributes to the cycle index the term $f_1{}^6$. Next let's try rotation about the center. One such rotation we've already looked at: a rotation of 60° ($2\pi/6$ radians).

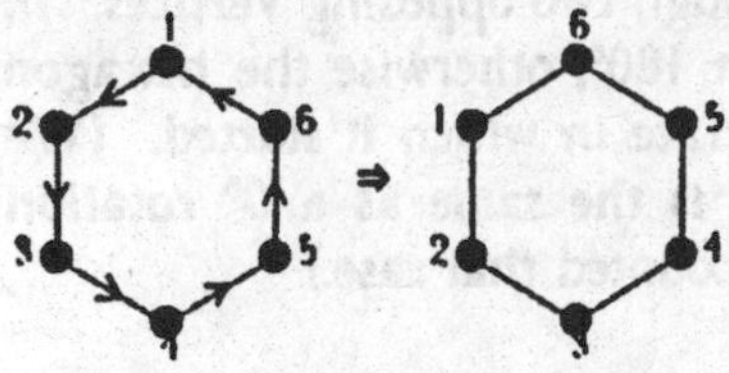

$$\Rightarrow \qquad \leftrightarrow \quad \begin{pmatrix} 1\ 2\ 3\ 4\ 5\ 6 \\ 2\ 3\ 4\ 5\ 6\ 1 \end{pmatrix} \quad \leftrightarrow \quad (1\ 2\ 3\ 4\ 5\ 6)$$

This permutation has one cycle of order 6. Meanwhile, there are two permutations of this form, because we could rotate the hexagon either counterclockwise (as shown) or clockwise. Together, these permutations contribute the term $2f_6$.

Moving on, we can rotate the hexagon 120° ($2\pi/3$ radians) about its center. This time the transposition of vertices does not take place along edges of the hexagon, so we will introduce dotted lines to indicate where the vertices end up.

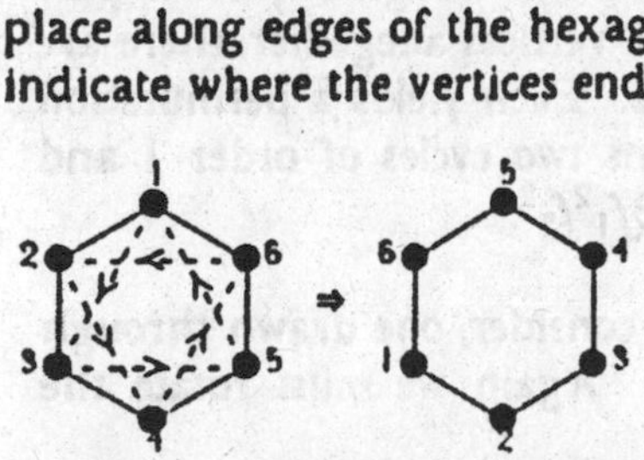

$$\Rightarrow \qquad \leftrightarrow \quad \begin{pmatrix} 1\ 2\ 3\ 4\ 5\ 6 \\ 3\ 4\ 5\ 6\ 1\ 2 \end{pmatrix} \quad \leftrightarrow \quad (1\ 3\ 5)(2\ 4\ 6)$$

As before, this rotation could be performed either clockwise or counterclockwise, so we get the term $2f_3{}^2$.

Finally, we could rotate the hexagon 180°. In this case, we don't get to choose between clockwise and counterclockwise, since rotation in either of the two directions yields the same permutation.

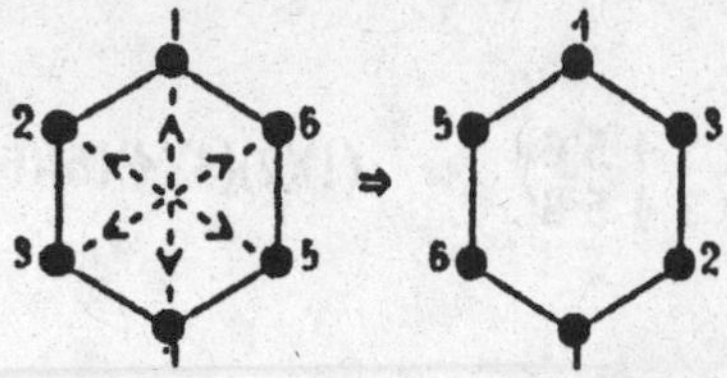

$$\Rightarrow \qquad \leftrightarrow \begin{pmatrix} 1 & 2 & 3 & 4 & 5 & 6 \\ 4 & 5 & 6 & 1 & 2 & 3 \end{pmatrix} \leftrightarrow (1\ 4)(2\ 5)(3\ 6)$$

Here we end up with three pairs of vertices being interchanged. Each such interchange is a cycle of order 2, so we get the term f_2^3.

So much for rotations about the center. We could also rotate the hexagon about an axis drawn through two opposing vertices. In this case the rotation must be through 180°; otherwise the hexagon will not end up in the same plane surface in which it started. (We could, of course, rotate it 0°, but this is the same as a 0° rotation about any other axis, so we've already counted that case.)

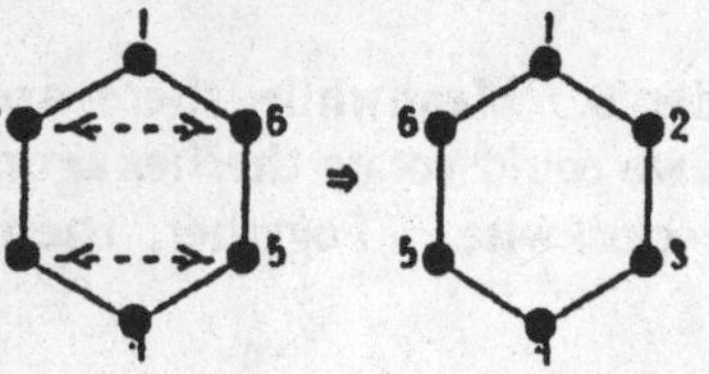

$$\Rightarrow \qquad \leftrightarrow \begin{pmatrix} 1 & 2 & 3 & 4 & 5 & 6 \\ 1 & 6 & 5 & 4 & 3 & 2 \end{pmatrix} \leftrightarrow (1)(2\ 6)(3\ 5)(4)$$

Here the axis was taken to be through the vertices labelled 1 and 4. We could just as easily have chosen it to be through any other pair of opposing vertices. Since there are six vertices altogether, there are three distinct pairs of opposing vertices. Each yields a permutation that, like the one shown above, contains two cycles of order 1 and two of order 2. Hence we get the term $3f_1^2f_2^2$.

There's one other type of axis to consider, one drawn through the midpoints of two opposing edges. Again we must rotate the hexagon through 180°.

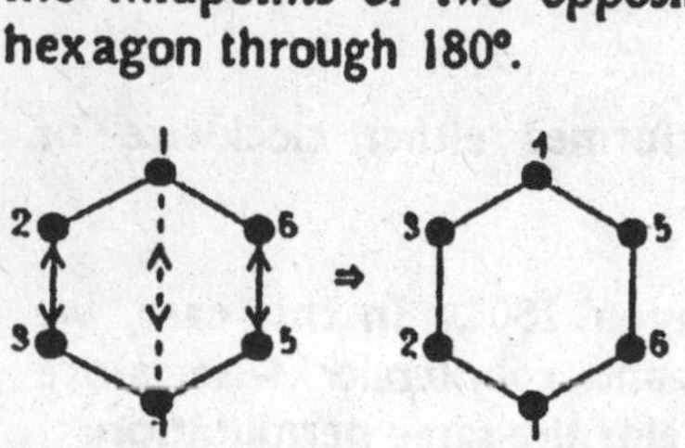

$$\Rightarrow \qquad \leftrightarrow \begin{pmatrix} 1 & 2 & 3 & 4 & 5 & 6 \\ 4 & 3 & 2 & 1 & 6 & 5 \end{pmatrix} \leftrightarrow (1\ 4)(2\ 3)(5\ 6)$$

Here the axis was taken to be through the edges labelled 2—3 and 5—6; we could have chosen any pair of opposing edges. Since there are six edges, there are three distinct pairs of opposing edges. Hence we get the term $3f_2^3$.

As a check, we add the coefficients of the terms we have generated for the cycle index. Since $1 + 2 + 2 + 1 + 3 + 3 = 12$, this accounts for all 12 different rotations. Adding everything together and dividing by 12 gives us the cycle index for this permutation group:

$$\frac{f_1^6 + 2f_6 + 2f_3^2 + f_2^3 + 3f_1^2f_2^2 + 3f_2^3}{12}.$$

Observe that the six operations involving rotation about the center also form a group, since any combination of those rotations yields another such rotation. This group is called a <u>subgroup</u> of the original group of 12 rotations. The group of order 12 is in turn a subgroup of the group that is formed by taking all 6! permutations of 6 elements.

Now, lest anybody suspect that none of this has any practical application, Pólya described an example out of organic chemistry. There is a chemical structure known as "benzene", consisting of six carbon atoms and six hydrogen atoms (chemical formula C_6H_6). Many different organic chemicals can be formed by substituting other atoms for one or more of the hydrogen atoms. In particular, consider the class of compounds called "twice-substituted benzene", C_6H_4XY, where X and Y represent the atoms that have taken the place of the two hydrogen atoms removed. As it happens, when two of the hydrogen atoms are replaced in this manner, it is possible to come up with any of three different compounds. These compounds have identical chemical formulas but different chemical properties. They are called isomerides or (the more modern term) "isomers" of each other. Chemists wondered why there are only three such isomers for any given atoms X and Y. They conjectured that it had something to do with the internal structure of the molecule.

One possibility, suggested by August Kekulé, was that the carbon atoms were arranged so as to form the vertices of a regular hexagon. Another suggestion was that they formed the vertices of an octahedron:

(You can visualise this figure in a number of ways. One way is to pretend the six vertices are positioned at the centers of the six faces of a cube. Another is to view it as two square pyramids glued together along their square faces.)

A third possibility, suggested by Albert Ladenburg, was that the carbon atoms formed the vertices of a triangular right prism, *i.e.*, a prism with its long faces perpendicular to the ends, and with ends in the shape of equilateral triangles.

The basic principle underlying each of these conjectures was that each of the carbon atoms should play the same rôle as any other; the vertices should not be distinguishable in any way.

How can we decide which model (if any) is right? We'll represent the X atom by a dark circle (●) and Y by a light circle (O). How many different configurations are there? In each model, the carbon atoms are indistinguishable, so it makes no difference where we attach the X atom. In how many distinct ways can we attach the Y? For the hexagonal model, there are three ways.

The other two positions for the Y atom are the same as the first two, merely flipped over. The octahedral model has only two ways to place the Y atom.

The other three positions are identical to the first, except rotated about a vertical axis.

The third model is a bit trickier. No two of the five possible positions for the Y atom can be rotated into each other without moving the X atom to a different position, so all five cases are different.

The reason this case is tricky is that the last two cases are exact "mirror images" of the first two. That is, if you reflect the last two in a mirror, they will come out looking exactly like the first two. The question is, are mirror images chemically distinct? The answer is: they are. Two chemicals that are mirror images of each other will have slightly different chemical properties. (In fact, if your body were reflected like this on the molecular level, your reflected body would not be capable of digesting ordinary food! Your food would have to be specially prepared to consist of mirror-image proteins and other reflected chemicals.)

What have we established? Well, if the carbon atoms were arranged as in an octahedron, there would be only two possible isomers of C_6H_4XY. If they were arranged as in a triangular prism, there would be five isomers. In reality, chemists were able to find only three; therefore these two models must be wrong. We haven't actually proven that the hexagonal model is correct, but we have circumstantial evidence in its favor. (Various other methods have backed this up, and modern chemists are essentially certain that the carbon atoms of benzene do indeed form a regular hexagon, at least insofar as chemical bonds maintain any rigid shape.)

This analysis, which enabled us to choose among the three

available models, was fairly easy. Suppose it had been $C_6H_2XYZ_2$, or some other exotic set of compounds? The intent of this chapter is to come up with a mathematical method for doing this sort of analysis.

The preceding computation by which we arrived at the cycle index for the rotations of the regular hexagon, can be summarised by the following chart.

Expected number of rotations: $6 \times 2 = 12$

Axis of rotation:	any	center	center	center	two edges	two vertices
Degree of rotation:	0°	±60°	±120°	180°	180°	180
Radians:	0	±2π/6	±2π/3	2π/2	2π/2	2π/2
Number of such axes:	1	1	1	1	3	3

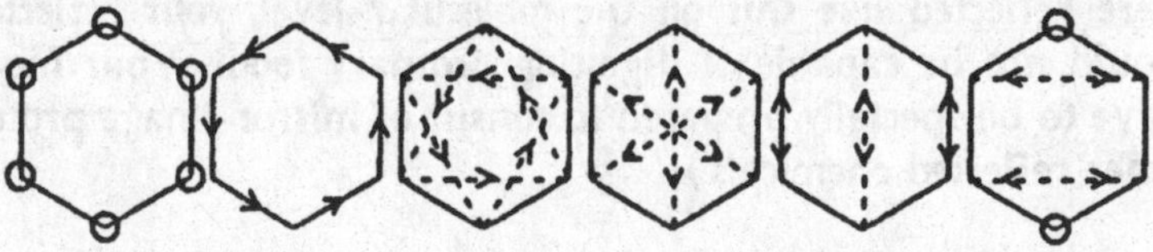

$$\frac{f_1^{6} \;+\; 2f_6 \;+\; 2f_3^{2} \;+\; f_2^{3} \;+\; 3f_2^{3} \;+\; 3f_1^{2}f_2^{2}}{12}$$

Let H be any permutation group. Let h be the order of H (the number of permutations in H), and let n be the degree of H (the number of objects being permuted). We shall define $h_{i_1,i_2,i_3...i_n}$ to be the number of permutations with i_k cycles of order k for all k between 1 and n. For example, if $n = 5$, then $h_{2,0,1,0,0}$ would be the number of permutations in H with 2 cycles of order 1, 1 cycle of order 3, and no other cycles. Note that, since each element being permuted must occur in one and only one cycle, we know that

$$1 \cdot i_1 + 2 \cdot i_2 + \cdots + n \cdot i_n$$

= total number of elements in all cycles

= n

if $h_{i_1 i_2 i_3 \dots i_n}$ is non-zero. Now, each permutation with i_1 cycles of order 1, i_2 cycles of order 2, i_3 of order 3, and so on, contributes to the cycle index the term

$$f_1^{i_1} f_2^{i_2} f_3^{i_3} \dots f_n^{i_n}.$$

Therefore the cycle index of H is

$$\frac{\Sigma h_{i_1 i_2 i_3 \dots i_n} \cdot f_1^{i_1} f_2^{i_2} f_3^{i_3} \dots f_n^{i_n}}{h},$$

where the summation is being performed over all combinations of non-negative integers $i_1, i_2, \ldots, i_n$ that satisfy the requirement that $1 i_1 + 2 \cdot i_2 + \cdots + n \cdot i_n = n$. Note, incidentally, that the total number of permutations is

$$\Sigma h_{i_1 i_2 i_3 \dots i_n} = h.$$

As homework, Pólya assigned the problem of finding the cycle index of the right equilateral triangular prism, where we consider the permutations of the vertices under rotations of the prism. He explicitly instructed that "mirror" reflections of the prism were not to be included. This was primarily because such reflections, combined with rotations, can yield some rather abstruse permutations, and the exercise was not intended to be one in solid geometry! Excluding reflections, we deduce that the group should be of order six, since there are six possible orientations for the prism. (We can choose any vertex to place at, say, the far left vertex of the upper face, and in so choosing we establish the position of all the other vertices as well. If mirror images were permitted, we would still have the choice of whether or not to reflect the prism through a plane running through the upper-left vertex and the middle of the opposing face, so there would be twelve permutations.) There are two types of axes we need to worry about. One is the axis running vertically through the centers of the two triangular faces. The others each run horizontally through the center of a vertical edge and the center of the opposing face.

There are three such axes. So, without further ado, here is the calculation of the cycle index for rotations of the prism.

Expected number of rotations: 6

Axis of rotation:	any	vertical	edge/face
Degree of rotation:	0°	±120°	180°
Radians:	0	±2π/3	2π/2
Number of such axes:	--	1	3

$$\frac{f_1^6 + 2f_3^2 + 3f_2^3}{6}$$

Just to see what we mean by "abstruse" rotations, let's look at what happens if we include reflections in the permutation group. We can reflect the prism across either a horizontal plane or a vertical plane, as shown at the top of page 67. There are three such vertical planes, but only one permissible horizontal plane (if the horizontal plane isn't exactly midway between the two triangular faces, the reflection won't cause the prism to coincide with itself), so we get four more permutations. But this gives us a total of only ten permutations, and we have already observed that there should be twelve. We cleverly deduce that we are missing two permutations, and after a bit of hunting we find that we can combine a reflection with a rotation to yield a new permutation. Specifically, if we reflect the prism through the horizontal plane, we can then rotate it about the vertical axis. This is best observed by numbering the vertices and watching what happens to them.

$$\rightarrow \quad \rightarrow \quad \leftrightarrow \begin{pmatrix} 1 & 2 & 3 & 4 & 5 & 6 \\ 5 & 6 & 4 & 2 & 3 & 1 \end{pmatrix} \leftrightarrow (1\,5\,3\,4\,2\,6)$$

There are two such permutations, depending on whether the rotation is performed clockwise or counterclockwise, so this accounts for the two "missing" cases. The complete cycle index for the triangular prism, including reflections as well as rotations, is

$$\frac{f_1^6 + 2f_3^2 + 3f_2^3 + f_2^3 + 3f_1^2f_2^2 + 2f_6}{12}.$$

It is interesting to note that this cycle index is the same as that for the rotations of a hexagon. This means that, if it were possible to reflect a prism, it would be in some sense indistinguishable from a hexagon with respect to the number of ways of marking their respective vertices. That is, if there are w distinct ways to mark the vertices of a hexagon under some criteria (*e.g.*, there are 3 ways to mark the vertices of a hexagon using one X atom and one Y atom), there must also be w ways to mark the vertices of a reflectable prism using the same criteria.

February 9. Moving on, it's finally time to see just why the cycle index is so important. (As Pólya put it, "The cycle index knows many things.") Suppose we have n "beads" (or colors of paint, or atoms, or whatever), of which r are red, s are silver, and t are tan ($r + s + t = n$). We wish to place these n beads at the vertices of some n-cornered figure, such as a regular n-sided polygon. We don't want to consider two arrangements as different if they can be transformed into one another by rotating the figure. In how many different ways can the n beads be placed at the n vertices?

We first find the cycle index for the permutation group H that consists of the permutations induced on the vertices by rotations of the figure. We then take the <u>figure inventory</u> $x + y + z$, in which x represents a red bead, y a silver bead, and z a tan bead. Here comes the important step: we *substitute the figure inventory into the cycle index* by replacing f_k with $x^k + y^k + z^k$. Note that this is *not* the usual algebraic interpretation of "substitution"! Finally, we expand the cycle index in powers of x, y, and z. (As we'll see, we don't necessarily need to compute *all* of the coefficients of this expansion.) The resulting coefficient of $x^r y^s z^t$ is the number of distinct ways of configuring r red, s silver, and t tan beads (with respect to the permutation group H).

Pólya did not take the time to prove that this indeed works, nor shall we do so in these notes. The proof is quite complicated and beyond the scope of this course. As Pólya phrased it, "The proof of the pudding is in the eating. You can't eat mathematics, but you *can* digest it." So let's "digest" this theorem by chewing on a few examples. For our first example we'll return to our old friend, the regular hexagon. Suppose we wish to attach one white circle, one black circle, and four gray circles to the vertices of the hexagon. If we think of the gray circles as being "invisible", we can see at once that this is the same situation we had earlier with the X and Y atoms. Therefore we know that there are exactly 3 distinct ways to attach the white and black circles (and in each case the remaining four vertices must have the gray circles). Let's check this using the cycle index.

We start with the cycle index from page 64. Previously we had deliberately refrained from combining similar terms, in order to preserve the correspondence between terms of the cycle index and rotations of the hexagon, but this time, to save computation, we'll combine such terms.

$$\frac{f_1^6 + 2f_6 + 2f_3^2 + 4f_2^3 + 3f_1^2f_2^2}{12}$$

Now we substitute the figure inventory $x + y + z$ as instructed, and get

$$\frac{(x+y+z)^6 + 2(x^6+y^6+z^6) + 2(x^3+y^3+z^3)^2 + 4(x^2+y^2+z^2)^3 + 3(x+y+z)^2(x^2+y^2+z^2)^2}{12}.$$

Expanding this expression into powers of x, y, and z is not a very tempting prospect. Fortunately, we needn't expand very much of it. Instead we can take advantage of an extension of the Binomial Theorem, to wit, the Multinomial Theorem (shown here for three variables—the generalisation to more than three variables should be obvious):

$$(x+y+z)^n = \sum_{r+s+t=n} \frac{n!}{r!s!t!} x^r y^s z^t.$$

So, for example, the coefficient of xyz^4 in $(x+y+z)^6$ is $\frac{6!}{1!1!4!}$ which equals 30. $2(x^6+y^6+z^6)$ has no term of the form $k \cdot xyz^4$, nor does the third term, nor the fourth. The last term is

$$3(x+y+z)^2(x^2+y^2+z^2)^2$$

$$= 3(x^2+y^2+z^2+2xy+2yz+2xz)(x^4+y^4+z^4+2x^2y^2+2y^2z^2+2x^2z^2).$$

The only xyz^4 term coming out of that comes from $2xy \cdot z^4$ which, multiplied by the outer coefficient 3, gives a coefficient of 6. So the coefficient of xyz^4 in the expanded cycle index (with the figure inventory substituted in) is just $30 + 6 = 36$. We now divide by the 12 and get a final answer of 3, which checks.

It was assigned as homework to use the cycle index of the prism to check the number of ways of placing one X atom and one Y atom thereon. We already know that the answer should be 5. Substituting the figure inventory into the cycle index shown on page 66, we get

$$\frac{(x+y+z)^6 + 2(x^3+y^3+z^3)^2 + 3(x^2+y^2+z^2)^3}{6}.$$

Once again we're looking for the coefficient of xyz^4. The first term yields a coefficient of 30 as before, and the other two expansions yield no xyz^4 terms. So the coefficient of xyz^4 in the expanded cycle index is 30, which when divided by 6 gives the expected answer, 5.

Another example: suppose we have three colors of paint, and we wish to paint the corners of an equilateral triangle. In how many ways can we do this? First, we find the cycle index. We computed this back on page 57, so we won't do it here. We substitute the figure inventory $x + y + z$, representing the three different colors of paint, into the cycle index to get

$$\frac{(x+y+z)^3 + 3(x+y+z)(x^2+y^2+z^2) + 2(x^3+y^3+z^3)}{6}.$$

This time we haven't specified which combination of colors we're interested in, so we'd best compute all the coefficients in the expansion. Even so, it's not as difficult as it looks, because the expression is symmetric with respect to x, y, and z. Thus it suffices to look at the coefficients of x^3, x^2y, and xyz. Every other term is symmetrically equivalent to one of these; for instance, y^2z has the same coefficient as x^2y. So we proceed to fill out a little table showing the coefficients of the three different kinds of terms in each part of the expansion.

	x^3	x^2y	xyz
$(x+y+z)^3$	$\frac{3!}{3!0!0!} = 1$	$\frac{3!}{2!1!0!} = 3$	$\frac{3!}{1!1!1!} = 6$
$3(x+y+z)(x^2+y^2+z^2)$	3	3	0
$2(x^3+y^3+z^3)$	2	0	0

Adding up each column and dividing through by 6, we find that every coefficient is 1. It's time for another of Pólya's maxims: *"If you see a fact, try to see it as intuitively as possible."* What does it mean for every coefficient to be 1? It means that each combination of colors can be applied in exactly one way. We'll see later why this should be the case.

The $n!$ possible permutations of n elements form a group, since if you reorder n objects once, and then reorder them again, you get yet another ordering. This group has degree n and order $n!$, and is called the <u>symmetric group</u>. We shall denote it by S_n. Let's try to find the cycle index of this group.

February 14. Consider any single permutation. Suppose it has i_1 cycles of order 1, i_2 of order 2, ..., and i_n of order n. The total number of elements in these cycles, as we've observed before, must be

$$1 \cdot i_1 + 2 \cdot i_2 + \cdots + n \cdot i_n = n .$$

What is $h_{i_1 i_2 i_3 ... i_n}$? (The following proof is due to Cauchy.) We consider any permutation, broken up into its component cycles. First we show the cycles of order 1, representing each by a square.

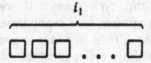

We then show the cycles of order 2, representing each by two squares.

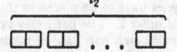

We carry on in this fashion until we've shown all the cycles.

Having put the permutation into this form, we note that there must be exactly n squares involved. So we proceed to write *any* permutation of the numbers 1 to n inside the squares. For instance, suppose $n = 9$, and we start with the permutation $\binom{1\,2\,3\,4\,5\,6\,7\,8\,9}{4\,7\,9\,1\,2\,6\,8\,5\,3}$. In the cycle notation this is written (6)(1 4)(3 9)(2 7 8 5), so our "square" drawing looks like this:

We then take an arbitrary permutation of the numbers 1 through 9 and plunk the numbers into the squares. For instance, let's use the permutation: 5 9 1 2 7 4 3 6 8.

$$\boxed{5}\quad \boxed{9}\,\boxed{1}\quad \boxed{2}\,\boxed{7}\quad \boxed{4}\,\boxed{3}\,\boxed{6}\,\boxed{8}$$

If we interpret this as a collection of cycles, we get the permutation (5)(9 1)(2 7)(4 3 6 8), which can also be written as $\left(\begin{smallmatrix} 1 & 2 & 3 & 4 & 5 & 6 & 7 & 8 & 9 \\ 9 & 7 & 6 & 3 & 5 & 8 & 2 & 4 & 1 \end{smallmatrix}\right)$. In other words, by taking permutations of the integers 1 through n and placing them in the squares, we can obtain a mapping from those permutations onto the set of permutations with this particular cycle form. Of course, more than one permutation, when placed within the squares, may yield the same result. For example, consider the permutation: 5 7 2 9 1 8 4 3 6. Placed in the squares, this produces

$$\boxed{5} \quad \boxed{7\,2} \quad \boxed{9\,1} \quad \boxed{8\,4\,3\,6}$$

which is once again the permutation $\left(\begin{smallmatrix} 1 & 2 & 3 & 4 & 5 & 6 & 7 & 8 & 9 \\ 9 & 7 & 6 & 3 & 5 & 8 & 2 & 4 & 1 \end{smallmatrix}\right)$. Consider any permutation P with this cycle form. If we plug each of the $n!$ permutations into the squares, how many times do we produce P? If P has i_1 cycles of order 1, then in order to produce P the i_1 elements that are contained in those cycles must be placed in the i_1 squares representing cycles of order 1. However, they can be placed in any order within those squares, so there are $i_1!$ ways to do it. Meanwhile, if P has i_2 cycles of order 2, then to produce P the i_2 pairs of elements in those cycles have to be placed in the i_2 pairs of squares representing cycles of order 2. The pairs may be placed in any of $i_2!$ orders, and furthermore each pair can be rotated; that is, the cycle $(a\ b)$ can also be written $(b\ a)$. Since there are i_2 cycles, each of which can be rotated into either of two forms, we get a factor of 2^{i_2}. In general, suppose P has i_k cycles of order k. Then these cycles can be arranged in any of $i_k!$ ways, and each cycle can be rotated into any of k forms, so we get a factor of $i_k!k^{i_k}$. Altogether, therefore, there are

$$i_1!1^{i_1} \cdot i_2!2^{i_2} \cdot i_3!3^{i_3} \cdot \ldots \cdot i_n!n^{i_n}$$

permutations that, when placed in the squares, yield the cycles of P. If we plug every one of the $n!$ possible permutations into the squares, each permutation with this particular cycle form is generated exactly that many times. But we "know" how many permutations have this form; at least, we've given it a name. So we conclude that

$$h_{i_1 i_2 i_3 \ldots i_n} \cdot i_1!1^{i_1} \cdot i_2!2^{i_2} \cdot i_3!3^{i_3} \cdot \ldots \cdot i_n!n^{i_n} = n!$$

and that the cycle index of the symmetric group S_n is therefore

$$\frac{\sum h_{i_1 i_2 i_3 \ldots i_n} \cdot f_1^{i_1} f_2^{i_2} f_3^{i_3} \ldots f_n^{i_n}}{r_i}$$

$$= \left(\frac{1}{n!}\right) \sum \frac{n!}{i_1! 1^{i_1} \cdot i_2! 2^{i_2} \cdot i_3! 3^{i_3} \cdot \ldots \cdot i_n! n^{i_n}} f_1^{i_1} f_2^{i_2} f_3^{i_3} \ldots f_n^{i_n}.$$

Recall that this summation is over all values of the i's such that $1 \, i_1 + 2 \cdot i_2 + \cdots + n \cdot i_n = n$. There is a trick we can use to find all such values. We observe that

$$n = 1 \, i_1 + 2 \cdot i_2 + \cdots + n \cdot i_n = \overbrace{1 + 1 + \cdots + 1}^{i_1} + \overbrace{2 + 2 + \cdots + 2}^{i_2} + \cdots.$$

In other words, each set of values for i_1 through i_n corresponds to a unique way of representing the number n as a sum of positive integers. So if we look at all sets of positive integers that sum to n, we can compute from them all possible values for the i's.

Let's see how this works by looking at the case $n = 3$. We can write 3 as $1+1+1$, which corresponds to the case $i_1 = 3$, $i_2 = 0$, $i_3 = 0$. We can also write 3 as $1+2$, or we could simply write it as 3. These cases are tabulated below.

	i_1	i_2	i_3	cycle form	cycle index coefficient
$3 = 1 + 1 + 1$	3	0	0	f_1^3	$\frac{3!}{3! 1^3} = 1$
$3 = 1 + 2$	1	1	0	$f_1 f_2$	$\frac{3!}{1! 1^1 \, 1! 2^1} = 3$
$3 = 3$	0	0	1	f_3	$\frac{3!}{1! 3^1} = 2$

(We've omitted factors of $0! k^0$ from the denominators in the last column.) So the cycle index of S_3 is

$$\frac{f_1^3 + 3 f_1 f_2 + 2 f_3}{6}$$

This should look familiar—it's the cycle index for rotations of a triangle. This explains why there's only one way to paint the vertices of a triangle with any given set of colors; if we try to apply the same colors to the vertices in some other order, we can always permute the vertices back to the first order, because no matter what the rotation is it must be in the symmetric group.

What about S_4? Does its cycle index correspond to anything tangible, such as rotations of a square or tetrahedron? Yes, but it's neither of those. The square has 8 rotations, while the tetrahedron has 12. The symmetric group, on the other hand, has $4! = 24$ permutations, so it can't correspond to either of those figures. Let's work it out and see what it looks like. We must first find all possible ways of representing 4 as a sum of positive integers. If the positive integers include a 1, then the remaining integers sum to 3, and we already know there are three cases. If the positive integers do not include a 1, then we can have 2+2, or simply 4. Here we go then with another table:

	ℓ_1	ℓ_2	ℓ_3	ℓ_4	cycle form	cycle index coefficient
4 = 1+1+1+1	4	0	0	0	f_1^4	$\frac{4!}{4!1^4}=1$
4 = 1+1+2	2	1	0	0	$f_1^2 f_2$	$\frac{4!}{2!1^2 1!2^1}=6$
4 = 1+3	1	0	1	0	$f_1 f_3$	$\frac{4!}{1!1^1 1!3^1}=8$
4 = 2+2	0	2	0	0	f_2^2	$\frac{4!}{2!2^2}=3$
4 = 4	0	0	0	1	f_4	$\frac{4!}{1!4^1}=6$

As a check, the coefficients should sum to 4!. $1 + 6 + 8 + 3 + 6 = 24 = 4!$. So S_4 has the cycle index

$$\frac{f_1^4 + 6f_1^2 f_2 + 8f_1 f_3 + 3f_2^2 + 6f_4}{24}.$$

It turns out that this is the cycle index for rotations of the *diagonals* of a cube, as was discovered on the midterm. (We'll have more to say about this in the solution to the midterm, which is in chapter 8.)

Now we come to the grand finale for this chapter, but don't

hold your breath—it's a *long* finale! So far we've only considered figure inventories of a rather limited form: the sum of distinct variables. When we had k possible choices from which to pick a "value" to be assigned to each of the elements being permuted, we associated a unique variable with each choice and took the figure inventory to be the sum of those variables. For example, if we were coloring the vertices of the triangle using three colors, we associated each color with a variable and took the sum: $x + y + z$. There is a more general form of the figure inventory, which we shall now examine.

The basic idea is that we associate with each choice a value that is not necessarily a unique variable. One of the suggested references ([Liu], pages 145-146) has a good example that introduces this concept; it is presented here with the permission of the author. Suppose you have three balls you wish to paint. You have three types of paint available—a cheap red paint, an expensive red paint, and a blue paint. In how many ways can you paint the balls? We wish to count as different the cases where the same paints are used but in different orders; that is, the case where the first two balls are painted blue and the last one cheap red is to be counted separately from the case where the first ball is painted cheap red and the other two blue, and so forth. Hence we are not allowing any permutations other than the identity permutation, and the cycle index is simply f_1^3.

We can represent the three kinds of paint by three variables, r_1 (expensive red), r_2 (cheap red), and b (blue). The figure inventory is $r_1 + r_2 + b$. Substituting this into the cycle index yields

$$(r_1 + r_2 + b)^3 = r_1^3 + r_2^3 + b^3 + 3r_1^2 r_2 + 3r_1 r_2^2 + 3r_1^2 b$$
$$+ 3r_2^2 b + 3r_1 b^2 + 3r_2 b^2 + 6r_1 r_2 b.$$

Thus, for instance, the term $3r_1 r_2^2$ means that there are three ways to paint the balls using the expensive red paint on one ball and the cheap red paint on the other two. Suppose, however, that we were to represent both the cheap red paint and the expensive red paint by the same variable, r. It is clear that the coefficient of, say, $r^2 b$, will be the sum of the coefficients of $r_1^2 b$, $r_1 r_2 b$, and $r_2^2 b$. In other words, the coefficient of $r^2 b$ will be the number of ways of painting the balls using *any* combination of the red paints on two balls and blue paint on the remaining ball.

$$(r + r + b)^3 = (2r + b)^3 = 8r^3 + 12r^2b + 6rb^2 + b^3$$

We observe, for example, that there are 8 ways to paint all three balls red. This makes sense, since for each ball we can choose either of the two red paints.

Let's take a closer look at what we've just done. Instead of saying that we had three kinds of paint, we said we had only two kinds, but that there were two paints of the first kind and one of the second. If we had had five kinds of red paint, three kinds of blue paint, and two kinds of green paint, we would have used the figure inventory $5r + 3b + 2g$. We can generalise this still further. Suppose we wish to place colored beads at the vertices of a regular hexagon. Each bead can be any of three colors, but there are different kinds of beads for each color. For instance, there may be round red beads, cubical red beads, and elongated red beads, and similarly let us suppose there are three kinds of blue beads and three kinds of green beads. Then we would use the figure inventory $3r + 3b + 3g$. If we were to substitute this into the cycle index, the coefficient of $r^i b^j g^k$ would be the number of ways of placing i red beads, j blue beads, and k green beads on the vertices of the hexagon, where two ways are considered different if they use differently shaped beads. This is essentially the same sort of thing as what we did with the different kinds of paint in the previous example. Now for that next step: Suppose, for simplicity, that there is only one shape for each color of bead, but that we are placing *clusters* of beads at each vertex. We'll assume each cluster must be one of a finite number of possibilities. For instance, we can place two red beads and a blue bead together at a vertex, or we can place three green beads, or we can place a blue bead and a green bead. No other combinations are allowed. Then we shall represent this by the figure inventory $r^2b + g^3 + bg$. If we substitute *this* figure inventory into the cycle index, the resulting coefficient of $r^i b^j g^k$ is the number of distinct ways of placing clusters of beads such that exactly i red beads, j blue beads, and k green beads are used.

The general form of the figure inventory is defined to be

$$f(x,y,z) = \sum_{r,s,t=0}^{\infty} a_{rst}x^r y^s z^t,$$

where a_{rst} is the number of figures with r "red beads", s "silver beads", and t "tan beads". Of course, these various colors of beads can actually be any identifying characteristics. It should be noted that the above definition is for the specific case of there being three different characteristics, represented by x, y, and z. Observe also that, if the inventory contains exactly one figure that has only a red bead, and one that has only a silver bead, and one that has only a tan bead, then the figure inventory degenerates into the special case $x + y + z$, which is where we started.

To substitute this general figure inventory into the cycle index, we replace f_1 in the index by $f(x,y,z)$, f_2 by $f(x^2,y^2,z^2)$, and so on. In general, f_k is replaced by

$$f(x^k,y^k,z^k) = \sum_{r,s,t=0}^{\infty} a_{rst}x^{kr} y^{ks} z^{kt}.$$

When we perform this substitution, the cycle index becomes

$$\sum_{r,s,t=0}^{\infty} A_{rst}x^r y^s z^t,$$

where A_{rst} is the number of configurations, <u>different with respect to the permutation group</u> whose cycle index we used, that have r "red beads", s "silver beads", and t "tan beads".

This is a fairly complex subject, so to illustrate it we have a fairly complex example. We turn once again to organic chemistry. This time we are concerned with the class of compounds known as "aliphatic alcohols". Alcohols, aliphatic and otherwise, have the chemical formula $C_nH_{2n+1}OH$, for various values of n; the term "aliphatic" means that the carbon atoms do not form a closed loop, as they do in benzene.

Knowing that a carbon atom has four valences and that the hydrogen atoms and the hydroxyl group (OH) have one apiece, we can draw some of the simpler aliphatic alcohols. We will represent a carbon atom by a diamond ($\Diamond$), a hydrogen atom by a circle (O), and the hydroxyl group by an arrow ($\downarrow$). If $n = 1$, we have only one molecule,

which happens to be methyl alcohol (also called methanol). If $n = 2$, we again have only one molecule. This time it's ethyl alcohol, which is the kind people drink. (Methanol, on the other hand, is highly poisonous.)

Note that this is indeed the same structure as

even though the two may look slightly different. If we were to represent the structures in three dimensions, the equivalence would be more apparent. We have to go to $n = 3$ to find an alcohol with two isomers.

(The one on the left is called n-propyl alcohol; the one on the right is called isopropyl alcohol. So much for today's lesson in organic chemistry.) The question to which we wish to address ourselves is, for any given n, how many different aliphatic isomers are there of $C_nH_{2n+1}OH$? We shall call this number R_n. The question may appear trivial at first glance, until we realise that the carbon atoms need not always form a straight chain. The following randomly selected alcohol for which $n = 8$ (2,4-dimethyl-2-hexanol, if you *must* know!) gives us some idea of the sort of complexity we're dealing with.

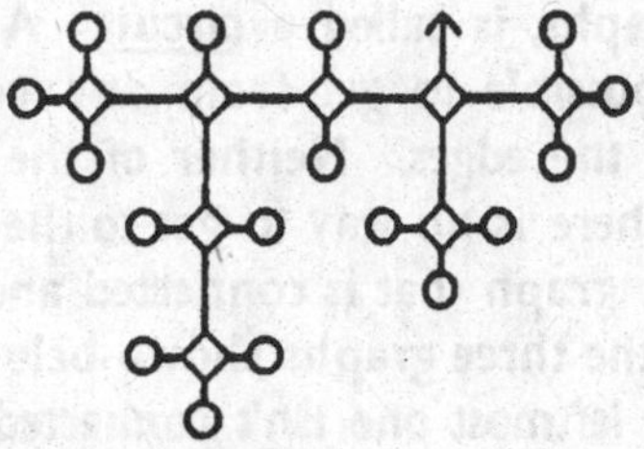

February 16. How can we get a handle on this problem? For starters, we can simplify the diagrams by omitting all the hydrogen atoms. For example, we'll draw *n*-propyl alcohol this way:

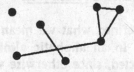

We can get away with doing this because we can always reconstruct the original form by realising that any unused "vertex" on a carbon atom must have a hydrogen atom attached.

This somewhat more compact form of the molecular structure can be thought of as a <u>rooted tree</u>. Now we need to back off and define some terms in order to understand what a rooted tree is. We'll start by defining a <u>graph</u>. A graph consists of a set of points, often called <u>vertices</u>, together with a set of <u>edges</u> connecting pairs of vertices. For example, the following graph has seven vertices (shown as dark circles) and five edges.

Note that not all of the vertices need be involved in edges. Note also that the crossed edges do *not* imply the existence of a vertex at the intersection. If we *do* put a vertex there we get a different graph, one with eight vertices and seven edges.

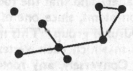

A sequence of edges forming a closed loop, such as the triangle in the preceding graphs, is called a <u>circuit</u>. A graph is said to be <u>connected</u> if it is possible to get from any vertex to every other vertex by following the edges. Neither of the preceding graphs is connected, because there is no way to get to the isolated vertex from any other vertex. A graph that is connected and that has no circuits is called a <u>tree</u>. Of the three graphs shown below, only the rightmost one is a tree. (The leftmost one isn't connected, whereas the center one has a circuit.)

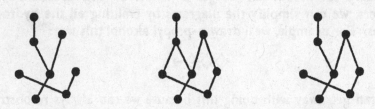

One noteworthy feature of trees, which you might try proving, is that in any tree the number of vertices is one greater than the number of edges. In a tree, an "endpoint" (that is, a vertex with only one edge) is called a <u>leaf</u>. In general, the number of edges entering a vertex is called the <u>order</u> or <u>degree</u> of that vertex. Thus the tree shown above has one vertex of degree 3, four of degree 2, and three of degree 1. (These last three are therefore leaves.) If one of the vertices in a tree is distinguished from the others in some manner (<i>e.g.</i>, by marking it with an arrow), that vertex is called the <u>root</u> of the tree, and the tree is called a <u>rooted tree</u>.

Now that we've defined what we mean by a rooted tree, let's get back to chemistry. In an aliphatic alcohol, the carbon atoms must certainly be connected, since otherwise we'd have two smaller molecules. Furthermore, by the definition of the term "aliphatic", there can be no closed loops. Hence the carbon atoms form a tree. We'll let the carbon atom that is connected to the hydroxyl group (OH) be the root of the tree. We observe that no carbon atom can be bonded to more than four other carbon atoms (or four of any kind of atoms, for that matter), and that the root can be connected to at most three other carbon atoms, since one of its valences is being used to bond with the hydroxyl group. This means that the carbon atoms form a rooted tree in which each vertex has degree ≤ 4 and the root has degree ≤ 3. Conversely, any rooted tree meeting these criteria regarding the degrees of the vertices corresponds to a unique aliphatic alcohol. We have therefore transformed our chemical problem into the mathematical one of determining the number of such rooted trees with a given number of vertices.

Chemists solved this problem by trial and error for some small values of n. Our job is to find the solution mathematically. As usual, we'll start by inspecting the simplest cases and looking for a pattern. We can create all the trees with n vertices by taking each tree with $n-1$ vertices and attaching another vertex in all possible ways. (Some n-vertex trees may be generated more than once, in which case we'll eliminate the duplicates.)

There's no immediately obvious pattern. We might guess that R_n is always a power of two, but we'd be wrong. As it turns out, R_5 *does* happen to be 8, but R_6 is 17. Though it's always a good idea to look for a pattern, patterns can sometimes be deceiving!

Our next approach is to attempt to apply recursion. Is there any way we can look at a rooted tree as a combination of smaller rooted trees? Suppose we look at an arbitrary rooted tree.

We take the root and remove it from the graph. This breaks the tree into, in this particular case, three smaller trees, called subtrees. We will consider the root of each subtree to be the vertex that was connected to the root of the original graph.

We can reconstruct the original tree by attaching each of the three roots to a new vertex, and making that vertex be the new root.

In the trees in which we're interested, the root can be adjacent to at most three other vertices. Therefore there are at most three subtrees. If the root is adjacent to fewer than three vertices, we can still break the graph into three subtrees by using a "null tree", *i.e.*, one with no vertices.

The order of the subtrees is unimportant. That is, we get the same tree if we attach the subtrees in some other order.

Therefore we consider trees resulting from such permutations of the subtrees to be identical. This gives us the permutation group S_3, the symmetric group.

Now we're ready to apply Pólya's Theory of Counting to this problem. We have only one kind of "bead" this time, namely the carbon atoms (which correspond to vertices in the trees). We'll represent a carbon atom by the variable x. Then a tree that has k carbon atoms is represented by x^k. Our figure inventory is

$$r(x) = \sum_{k=0}^{\infty} R_k x^k$$

(since there are, by definition, R_k rooted trees with k vertices that satisfy the requirements regarding the maximum degrees of vertices). We let $R_0 = 1$ to account for the one form of null subtree. We substitute the figure inventory into the cycle index for S_3 and get

$$\frac{r(x)^3 + 3r(x)r(x^2) + 2r(x^3)}{6}.$$

The coefficient of x^t in this formula is the number of ways, distinct under the permutation group S_3, of choosing three subtrees to connect to the root of a new tree, such that the total number of carbon atoms in the three subtrees is t. Each tree resulting from such a selection will actually have $t+1$ carbon atoms, the extra atom being the newly added root. So if we multiply the above formula by x, the coefficient of x^{t+1} will be the number of distinct rooted trees with $t+1$ vertices. But this is simply $r(x)$. Well, not quite. The coefficient of x^{t+1} is the number of such trees only for $t \geq 0$. We need to add the appropriate coefficient for x^0 explicitly. This coefficient is $R_0 = 1$. Thus we conclude that

$$1 + x\left(\frac{r(x)^3 + 3r(x)r(x^2) + 2r(x^3)}{6}\right) = r(x).$$

We have derived a _functional equation_ for $r(x)$. No one has ever managed to solve it to produce an explicit formula for $r(x)$, but it is nevertheless possible to use it to compute R_n by expanding the formula in powers of x. Due to the factor of x on the left-hand side, the coefficient of x^n on the left can never involve any coefficients of x^m for $m \geq n$. For example, we know that

$$r(x) = 1 + x + x^2 + 2x^3 + 4x^4 + R_5x^5 + R_6x^6 + \cdots.$$

Suppose we pretend that the "$\cdots$" isn't there, *i.e.*, that there are no further terms. If we expand the left-hand side of the functional equation (credit is due to MIT's MACSYMA system for aid in doing this expansion), we find

$$r(x) = 1 + x + x^2 + 2x^3 + 4x^4 + R_5x^5 + R_6x^6$$

$$= 1 + x\left(\frac{r(x)^3 + 3r(x)r(x^2) + 2r(x^3)}{6}\right)$$

$$= 1 + x + x^2 + 2x^3 + 4x^4 + 8x^5 + (R_5+9)x^6$$
$$+ (R_6+R_5+14)x^7 + (R_6+2R_5+17)x^8$$
$$+ (2R_6+3R_5+25)x^9 + (3R_6+7R_5+22)x^{10} + \cdots$$
$$+ \frac{R_5R_6{}^2+R_5R_6}{2}x^{18} + \frac{R_6{}^3+3R_6{}^2+2R_6}{6}x^{19}.$$

Since for two power series to be equal they must have equal coefficients for all pairs of corresponding terms, we find that $R_5 = 8$, and that $R_6 = R_5 + 9 = 17$. Note that, in the expanded formula, the coefficient of x^5 did not depend on R_5, and that the coefficient of x^6 did not depend on R_6. This continues to be true for later terms. Thus the coefficient of x^7 does not depend on the fact that our partial value for $r(x)$ had a zero coefficient for x^7, and we can conclude that $R_7 = 39$. The remaining terms in the expansion are of no use to us, since they *do* depend on the coefficient of x^7 in $r(x)$. Indeed, if we were doing this by hand (MACSYMA, in case you don't recognise the name, is a computerised system for doing sophisticated symbolic mathematical manipulations), we wouldn't have bothered computing any terms beyond x^7. Having thus found R_5, R_6, and R_7, we could extend our partial power series for $r(x)$, re-expand the functional equation, and evaluate a few more coefficients.

That's all we have to say about Pólya's Theory of Counting. For further reading, just about any good book on combinatorics will suffice, though most make use of rather complicated mathematical notations. Chapter 5 of [Liu] is probably as good as any other. The analysis involving aliphatic alcohols is included in the December, 1956, issue of the *American Mathematical Monthly*. For the more avid reader, chapters 4 and 5 of [Balaban] discuss methods, among them Pólya's theory, for counting various acyclic and cyclic chemical compounds, including unsubstituted alkanes (C_nH_{2n+2} as opposed to $C_nH_{2n+1}OH$), stereo-isomers (in which mirror images are counted as distinct molecules), and other complex structures.

7 | Outlook

Everything we've been doing in the preceding chapters has been counting combinatorics, that is, combinatorics in which we wish to compute the number of configurations meeting certain criteria. Such problems make up one branch of combinatorics; there are two other classes of problems on which we haven't even touched. In this chapter we look briefly at these other branches of combinatorics. Some of the subjects introduced here will be covered in greater depth by Tarjan in later chapters.

One of the other branches is existential combinatorics. In existential problems we no longer wish to count anything; we just want to know whether there exists *any* configuration meeting certain criteria. For example, suppose there are n people at a party. We'll assume n is at least 2 (otherwise it's not much of a party). Some people are acquainted with others, where "being acquainted" is a mutual relationship. That is, if a person x is acquainted with another person y, then y is likewise acquainted with x. Then we claim that *some two people* at the party have exactly the same number of acquaintances at the party. We don't know who the two people are; we don't know how many acquaintances they have; we don't know whether there are more than two people with the same number of acquaintances. But we claim that some two people have the same number. This sort of claim is in the domain of existential combinatorics.

The preceding claim is easy enough to prove. Each person at the party has some number of acquaintances. No person can have fewer than zero acquaintances, and no person can have more than $n-1$ acquaintances at the party. (We're assuming the relationship of "acquaintance with" does not apply to oneself. If we instead assume that everyone is acquainted with himself, it just adds 1 to each person's number of acquaintances, and so does not affect the result.) So there are exactly n possibilities for the number of acquaintances a person has. Suppose the claim is wrong; suppose no two of the n people have the same number of acquaintances. Then there must be exactly one person with no acquaintances, one person with one acquaintance, one with two, . . ., and one with $n-1$ acquaintances. This person who has $n-1$ acquaintances must be acquainted with

G. Pólya et al., *Notes on Introductory Combinatorics*, Modern Birkhäuser Classics,
DOI 10.1007/978-0-8176-4953-1_7, © Birkhäuser Boston, a part of Springer
Science+Business Media, LLC 2010

each of the other $n-1$ people. But one of those people is supposed to have no acquaintances. This is a contradiction; hence the claim must be correct.

The third area of combinatorics is <u>constructive</u> combinatorics. This area deals with problems where we don't care how many configurations exist meeting some criterion, nor whether they exist (usually it's obvious from the statement of the problem that *some* configuration exists); rather, we wish to *find* one such configuration. For example, suppose you own a number of shops in San Francisco. Each shop is run by one employee, who lives somewhere in the city. There are various bus routes running through the city, such that for each employee there are certain shops that he or she can reach directly by bus. You wish to assign your employees to the shops in such a way that as many of the employees as possible are able to go to work by buses. It may not be possible to assign the shops such that *all* the employees can use the buses, but it's clear that there must be at least one assignment that achieves the maximum possible, so this isn't a question of existential combinatorics. Similarly, we don't care how many assignments there are that achieve the maximum, so this isn't "counting" combinatorics. We just want to *construct* one solution.

In constructive combinatorics, the problem is usually one of finding a solution *efficiently*. In the busing problem just discussed, we could obviously find a solution by looking at each of the possible assignments and computing for each the number of employees who are able to use buses. On the other hand, since for n employees and n shops the number of possible assignments is $n!$, this could take a while. In chapter 11 we'll look at a way of solving this problem using a reasonable length of time.

February 21. Returning to existential combinatorics, Pólya next gave an example out of Ramsey Theory. Consider a graph in which each edge has been colored using one of two colors. (Because of the limitations of the photoreproductive process, we'll use solid and dotted lines to represent the two colors in these notes.)

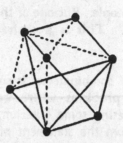

Having thus colored the edges, we find that the graph may or may not contain a triangle all three of whose edges are the same color. Such a triangle is called a <u>monochromatic</u> triangle. The above graph, colored as shown, does happen to include a monochromatic triangle (in fact, it contains two of them, both consisting of "solid" edges). This need not always be the case. For instance, the following five-vertex graph contains no such triangle.

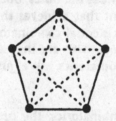

Notice that this graph has an edge between every pair of vertices. Such a graph is called a <u>complete graph</u>, and is denoted by K_n, where n is the number of vertices. (Quick question: how many edges are in K_n?) Ramsey Theory tells us that, for any $n \geq 6$, if we color the edges of K_n using two colors the resulting graph must contain a monochromatic triangle.

To prove this, we first observe that, if it's true for $n = 6$, it must be true for all $n > 6$. Given K_n, we can choose any 6 vertices. These vertices must be completely interconnected, so they form a K_6. If this K_6 contains a monochromatic triangle, then so does the larger graph. So how do we show that K_6 must contain such a triangle? We examine an arbitrary vertex in K_6. The vertex must have five edges coming out of it, one leading to each of the remaining vertices. Of these five edges, at least three must be the same color (as each other, that is). If this weren't true, then there could be at most two solid edges and at most two dotted edges. But this gives us at most

four edges, and there are supposed to be five. So we have either
three solid edges or three dotted edges. (We could have more than
three, but we don't care about that.) Without loss of generality, let's
assume the three edges are all solid.

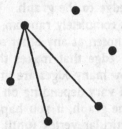

Now consider the three vertices at the "other ends" of those
edges. There must be edges between every pair of vertices, since this
is a complete graph. If any two of these three vertices are connected
by a solid edge, we've got a solid triangle. If no two of the three are
connected by a solid edge, then they must be connected by dotted
edges, and we've got a dotted triangle as shown below.

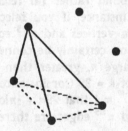

Notice that a triangle is simply a complete graph with three
vertices. Suppose we wished to prove that, for some sufficiently
large n, K_n cannot be colored with two colors without producing a
monochromatic K_4? It's not too difficult to prove that, for any s,
there is some number n sufficiently large that K_n must, if colored
with two colors, contain a monochromatic K_s. Unfortunately, no
general method is known for *finding* that number! We'll look at this
and related problems in chapter 9.

Existential combinatorics needn't always deal with conclusions
of the form, "Such-and-such exists." Sometimes the result can be
probabilistic. For example, Erdős has proved several results having
to do with randomly-generated graphs. We won't attempt to prove

any of these results here, but will simply mention one such result as an example of this sort of problem. Suppose you generate a graph by the following method. You start with n vertices and no edges. You then repeatedly pick any two vertices that are not yet joined by an edge, and add that edge to the graph. Your selection of which edge to add each time is completely random; any edge not yet in the graph is as likely to be chosen as any other edge. Stop adding edges as soon as you add an edge that makes the graph <u>connected</u> (as defined on page 80). How many edges are in the graph? Obviously, the number of edges will vary depending on the order in which you choose to add them to the graph; if you happen never to choose an edge involving some particular vertex (until, of course, there are no other edges from which to choose), then there will be many edges indeed. On the other hand, the graph could be connected after as few as $n-1$ edges have been chosen (in which case the graph would be a tree). Erdős has proved that, for any constant k, the probability that the graph has fewer than $kn + \frac{1}{2}n\log_e n$ edges approaches $e^{-e^{-k}}$ as n goes to infinity.

This result may sound rather far-fetched, but it has some simple corollaries. For instance, if you select a graph at random from among those with n vertices and $n^{1.5}$ edges, where n is fairly large, the graph will almost certainly be connected. This is because $n^{1.5}$ is, for sufficiently large n, greater than $kn + \frac{1}{2}n\log_e n$ for any constant k. So we can let $k = 99$ (for instance) and deduce that the probability is roughly $e^{-e^{-99}} \approx 1$ that $99n + \frac{1}{2}n\log_e n$ edges are enough to connect the graph, and $n^{1.5}$ edges are therefore many more than enough.

Incidentally, it should by no means be believed that we have exhausted the topics involving "counting" combinatorics. There are all sorts of things we've never even mentioned. Here, to close out this chapter, are two of them.

Consider subsets of some given set S. Some of these subsets are in turn subsets of other subsets of S. For instance, if S is the set $\{1,2,3,4,5,6\}$, then the set $\{1,3\}$, in addition to being a subset of S, is also a subset of $\{1,3,4,5\}$, which is in turn a subset of S. We'll say two subsets are <u>connected</u> if either is a subset of the other, and <u>disconnected</u> otherwise (even though they may have *some* elements in common). For example, $\{1,3\}$ and $\{1,3,4,5\}$ are connected, whereas

{1,3,6} and {1,3,4,5} are disconnected. (Note that this has nothing to do with the term "connected" as used with respect to a graph.) How many mutually disconnected subsets can we find in a set of size n?

For starters, we know that the $\binom{n}{k}$ subsets of size k are all mutually disconnected, since no subset of size k can possibly contain every element from another subset of size k. For what value of k is $\binom{n}{k}$ greatest? The answer is intuitively obvious from an examination of Pascal's triangle, but let's prove it rigorously. We know that

$$\binom{n}{k} = \frac{n!}{k!(n-k)!} = \frac{n-k+1}{k} \cdot \frac{n!}{(k-1)!(n-k+1)!}$$

$$= \frac{n-k+1}{k}\binom{n}{k-1}.$$

Hence $\binom{n}{k}$ will be greater than $\binom{n}{k-1}$ if $n-k+1$ is greater than k, which is true only if $k < \frac{1}{2}(n-1)$. So the maximum occurs when $k = \lfloor \frac{1}{2}(n-1)\rfloor$. If n is even, then, we let $n = 2m$ and find that the maximum is $\binom{2m}{m}$. If n is odd, we let $n = 2m+1$, and find that the maximum occurs twice: $\binom{2m+1}{m}$ and $\binom{2m+1}{m+1}$.

We've found the maximum possible value for $\binom{n}{k}$, but is it possible to find a larger collection of disconnected subsets by letting them be of varying sizes? It turns out, though it's difficult to prove and we won't attempt to do so here, that it's not. The maximum value for $\binom{n}{k}$ is indeed equal to the maximum number of mutually disconnected subsets.

Finally, let us consider the problem of counting all rooted trees that have n vertices. This is similar to the problem with which we closed the previous chapter, but in this case we no longer wish to place any restrictions on the degrees of the vertices. Because of this, we can no longer apply the method used in chapter 6. We would need to know the permutation group for the subtrees, but there can be any number of subtrees. Hence no symmetric group S_k is good enough, since there could be more than k subtrees. Instead, we'll use an approach bearing a strong resemblance to a generating function.

We wish to construct a rooted tree out of rooted subtrees. Which subtrees shall we use? Well, there's the subtree consisting of a single vertex. We could use it no times, or once, or twice, or thrice, etc. We'll represent these choices by a sum, using "1" to represent the case of no such subtree.

$$1 + \diamondsuit + \diamondsuit\,\diamondsuit + \diamondsuit\,\diamondsuit\,\diamondsuit + \cdots$$

Similarly, we could have no subtrees with two vertices, or we could have one, or two, etc.

$$1 + \diamondsuit + \diamondsuit\,\diamondsuit + \diamondsuit\,\diamondsuit\,\diamondsuit + \cdots$$

We have similar choices for how many times to use *each* of the two subtrees with three vertices.

$$1 + \diamondsuit + \diamondsuit\,\diamondsuit + \diamondsuit\,\diamondsuit\,\diamondsuit + \cdots$$

$$1 + \diamondsuit + \diamondsuit\,\diamondsuit + \diamondsuit\,\diamondsuit\,\diamondsuit + \cdots$$

We do this for each of the infinity of possible subtrees. Then we take the product of all these sums. Each term of the product will correspond to a selection of one term from each of the sums, and hence to a unique rooted tree, namely the one that has the selected terms as its subtrees.

If we substitute x^k in place of every tree of size k, we get

$$
\begin{aligned}
(1 &+ x + x^2 + x^3 + \cdots) \\
&\cdot (1 + x^2 + x^4 + x^6 + \cdots) \\
&\cdot (1 + x^3 + x^6 + x^9 + \cdots) \\
&\cdot (1 + x^3 + x^6 + x^9 + \cdots) \\
&\qquad\vdots
\end{aligned}
$$

Each factor in this infinite product will have the form

$$(1 + x^k + x^{2k} + x^{3k} + \cdots).$$

Notice that the factor for $k = 3$ occurs twice—once for each of the two possible subtrees of size 3. How many times does the factor for $k = s$ occur? It occurs as many times as there are subtrees of size s. Let's call that number T_s. Our infinite product of infinite sums is thus

$$(1 + x + x^2 + \cdots)^{T_1}$$
$$\cdot (1 + x^2 + x^4 + \cdots)^{T_2}$$
$$\cdot (1 + x^3 + x^6 + \cdots)^{T_3}$$
$$\vdots$$
$$\cdot (1 + x^k + x^{2k} + \cdots)^{T_k}$$
$$\vdots$$

which we can write more simply as

$$(1-x)^{-T_1}(1-x^2)^{-T_2}(1-x^3)^{-T_3} \cdots (1-x^k)^{-T_k} \cdots.$$

The coefficient of x^t in this product is the number of ways to choose subtrees that have a total of t vertices. Thus this coefficient is also the number of rooted trees that have $t+1$ vertices, since any combination of subtrees totalling t vertices can be put together to form a unique tree with $t+1$ vertices, the additional vertex being the new root. As in the aliphatic alcohol problem, we multiply by x and conclude

$$x(1-x)^{-T_1}(1-x^2)^{-T_2}(1-x^3)^{-T_3} \cdots (1-x^k)^{-T_k} \cdots$$
$$= T_1 x + T_2 x^2 + T_3 x^3 + \cdots + T_k x^k + \cdots.$$

(We could also add one to the left-hand side and add T_0 on the right, but that isn't particularly important.) As in the case of the aliphatic alcohols, we can now plug in the values of the first few T's, expand the left-hand side in powers of x, and thereby compute the next few T's. Plugging them in, we can then re-expand the power series and compute a few more, and so forth.

Actually, there is nothing to prevent us from computing any (finite) number of additional coefficients after each new expansion; we need only treat the unknown T_k as variables and solve for them after the expansion. For example, we can let T_1 through T_4 be variables, pretending that T_k is zero for all $k > 4$. We find that

$$x(1-x)^{-T_1}(1-x^2)^{-T_2}(1-x^3)^{-T_3}(1-x^4)^{-T_4}$$

$$= x + T_1 x^2 + (T_2 + \tfrac{1}{2}T_1^2 + \tfrac{1}{2}T_1)x^3 + \frac{6T_3 + 6T_1 T_2 + T_1^3 + 3T_1^2 + 2T_1}{6}x^4$$

$$+ \frac{24T_4 + 24T_1 T_3 + 12T_2(T_2 + T_1^2 + T_1 + 1) + T_1^4 + 6T_1^3 + 11T_1^2 + 6T_1}{24}x^5$$

Since the coefficient of x does not involve any T's, we can deduce that $T_1 = 1$. From that we can compute the coefficient of x^2, namely 1, and then $T_3 = 1 + \tfrac{1}{2} + \tfrac{1}{2} = 2$, $T_4 = (6 \cdot 2 + 6 + 1 + 3 + 2)/6 = 4$, and (since we can show that the coefficient of x^5 is independent of our assumption that T_5 was zero) $T_5 = [24 \cdot 4 + 24 \cdot 2 + 12 \cdot (1+1+1+1) + 1 + 6 + 11 + 6]/24 = 9$. For the curious, we turn once again to MIT's MACSYMA and find that the first several terms of the expansion are

$$1 + x + 2x^2 + 4x^3 + 9x^4 + 20x^5 + 48x^6 + 115x^7$$
$$+ 286x^8 + 719x^9 + 1842x^{10} + 4766x^{11} + \cdots,$$

where by comparison the generating function for the number of aliphatic alcohols begins

$$1 + x + 2x^2 + 4x^3 + 8x^4 + 17x^5 + 39x^6 + 89x^7$$
$$+ 211x^8 + 507x^9 + 1238x^{10} + 3057x^{11} + \cdots.$$

    So much for our "outlook", and so much for Pólya's portion of the course. Starting with chapter 9 we will be discussing material presented by Tarjan.

8 | Midterm Examination

This chapter contains the midterm exam and the solutions thereto. The exam was open book and "take home"; students were given one week to work on it. They were advised they would find the exam somewhat "open-ended". They were not required to do problems 1c and 2b, though they were strongly encouraged to do so. It was stated that extra credit would be given to students who attempted those problems.

At the time the midterm was presented, course notes had not yet been prepared for chapters 6 and 7, though the material had been covered in class. The table on page 64, summarising the computation of the cycle index for the hexagonal permutation group, was included in the midterm handout for reference.

The following note by Pólya was included at the end of the instructions for the exam, and so well epitomises what is desired from students taking *any* exam that we must include it here as well:

> *Good presentation counts!* It should be correct, complete, concise, and clear.

Problem 1 (50 points total).

The rotations that transform a cube into itself form a group (usually called the *octahedral group*; see below). The rotations of this group permute
  (1) the vertices of the cube,
  (2) the midpoints of the faces of the cube (vertices of an octahedron), and
  (3) the diagonals of the cube,
and so generate three permutation groups, of degrees 8, 6, and 4, respectively. (Note: By "diagonal" is meant a three-dimensional or "space" diagonal, which connects two opposite vertices. No single face of the cube touches both endpoints of such a diagonal.)

Part 'a' (30 points) Write down the cycle indices of the three permutation groups, combined with further appropriate data (such

G. Pólya et al., *Notes on Introductory Combinatorics*, Modern Birkhäuser Classics,
DOI 10.1007/978-0-8176-4953-1_8, © Birkhäuser Boston, a part of Springer
Science+Business Media, LLC 2010

as the axes and degrees of rotation) in a well-planned format so that the relations of the three groups to each other and to the figure (faces, vertices, and edges of the cube, or of the octahedron) are easily visible.

Part 'b' (20 points). "Substitute" into the three cycle indices the "figure inventory"

$$x + y + z$$

(representing three beads of different colors) and verify some of the resulting combinatorial numbers "by inspection".

Part 'c' (extra credit). Note any pertinent remarks (on the relations displayed in (a), on matters treated—or not treated—during the course, observed patterns, guesses, possibly proofs, etc.)

Problem 2 (20 points total).

Part 'a' (20 points). Prove the useful formula,

$$\sum_{k=0}^{n} \binom{k}{m} = \binom{0}{m} + \binom{1}{m} + \cdots + \binom{m}{m} + \cdots + \binom{n}{m} = \binom{n+1}{m+1}.$$

Part 'b' (extra credit). Using (a), find a formula for

$$\sum_{k=0}^{n} k^t = 0^t + 1^t + 2^t + \cdots + n^t.$$

Your formula should be a polynomial in n, or perhaps the sum of t such polynomials. (You may already know the formulas for $t=1$ and $t=2$. The one for $t=3$ is also worth noting.)

Problem 3 (30 points total).

Consider the decimal integers that have n digits, for $n \geq 2$. For example, for $n = 3$, the numbers are 100 through 999. (We are not allowing "leading zeroes", as in 007.) As a function of n, how many such numbers have no two adjacent digits alike? *I.e.*, we wish to include numbers like 747 but not 344. Prove your answer two ways—with and without using PIE.

SOLUTIONS

Problem 1 (50 points total).

Part 'a' (30 points). Regardless of what part of the cube we are looking at, be it the vertices, the faces, or the diagonals, there are 24 possible positions for the cube and hence 24 permutations in the group. We can see that there are 24 positions by observing that there are six faces, any of which can be positioned at the top, and that for each choice there are then four faces any of which can be positioned at the "front". The 24 rotations involve three kinds of axes. The axis can go through the centers of two opposing faces, or through the centers of two opposing edges, or through two opposing vertices. (In this last case the axis is a diagonal of the cube.) We know that these are the only axes we need to consider because the rotations about these axes yield 24 permutations, which are all we expected to find. Here is the table showing the derivation of the cycle indices of the three groups. In the interests of legibility, only the vertices' permutations are included in the diagrams.

Expected number of rotations: $6 \times 4 = 24$

Axis:	any	two faces	two faces	two edges	diagonal
Degrees:	0°	±90°	180°	180°	±120°
Radians:	0	±2π/4	2π/2	2π/2	±2π/3
axes:	--	3	3	6	4

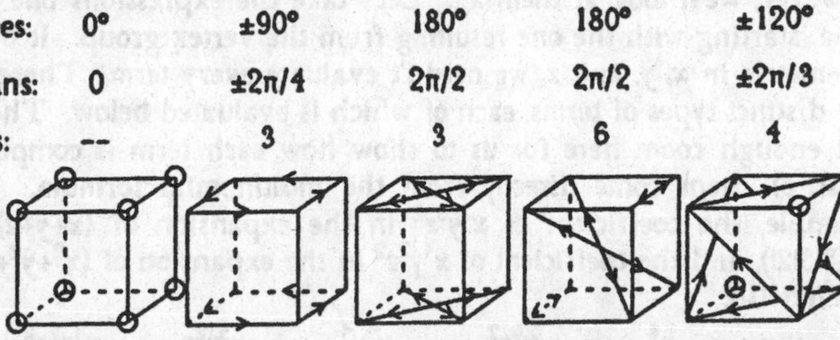

Vertices' cycle index

$$\frac{f_1^8 + 6f_4^2 + 3f_2^4 + 6f_2^4 + 8f_1^2 f_3^2}{24}$$

Faces' cycle index

$$\frac{f_1^6 + 6f_1^2 f_4 + 3f_1^2 f_2^2 + 6f_2^3 + 8f_3^2}{24}$$

Diagonals' cycle index

$$\frac{f_1^4 + 6f_4 + 3f_2^2 + 6f_1^2 f_2 + 8f_1 f_3}{24}$$

The cycle index for the vertices contains a pair of terms (the third and fourth) that could be combined, but it's probably better to leave them separate so we can better observe the similarities (and differences) among the three cycle indices. We can always combine the terms later when we are ready to substitute the figure inventory into the cycle index.

Part 'b' (20 points). Substituting the figure inventory $x + y + z$ into the three cycle indices yields the three expressions,

$$[(x+y+z)^8 + 6(x^4+y^4+z^4)^2 + 3(x^2+y^2+z^2)^4$$
$$+ 6(x^2+y^2+z^2)^4 + 8(x+y+z)^2(x^3+y^3+z^3)^2] / 24 ,$$

$$[(x+y+z)^6 + 6(x+y+z)^2(x^4+y^4+z^4) + 3(x+y+z)^2(x^2+y^2+z^2)^2$$
$$+ 6(x^2+y^2+z^2)^3 + 8(x^3+y^3+z^3)^2] / 24 , \text{ and}$$

$$[(x+y+z)^4 + 6(x^4+y^4+z^4) + 3(x^2+y^2+z^2)^2$$
$$+ 6(x+y+z)^2(x^2+y^2+z^2) + 8(x+y+z)(x^3+y^3+z^3)] / 24 .$$

It wasn't necessary on the exam to find the coefficients of *all* the terms in the power series expansions of these expressions, though many people did so. Determining a few typical coefficients and verifying the values "by inspection" was considered sufficient. Here, however, we'll look at them all. Let's take the expressions one at a time, starting with the one resulting from the vertex group. It being symmetric in x, y, and z, we needn't evaluate every term. There are ten distinct types of terms, each of which is evaluated below. There's not enough room here for us to show how each term is computed; most of them come directly from the multinomial formula. For example, the coefficient of $x^3y^3z^2$ in the expansion of $(x+y+z)^8$ is $8!/(3!3!2!)$, and the coefficient of $x^4y^2z^2$ in the expansion of $(x^2+y^2+z^2)^4$ is $4!/(2!1!1!)$.

	x^8	x^7y	x^6y^2	x^5y^3	x^4y^4	x^6yz	x^5y^2z	x^4y^3z	$x^4y^2z^2$	$x^3y^3z^2$
$(x+y+z)^8$	1	8	28	56	70	56	168	280	420	560
$6(x^4+y^4+z^4)^2$	6·1	0	0	0	6·2	0	0	0	0	0
$9(x^2+y^2+z^2)^4$	9·1	0	9·4	0	9·6	0	0	0	9·12	0
$8(x+y+z)^2 \dots$	8·1	8·2	8·1	8·2	8·4	8·2	0	8·4	0	8 2
Total ÷ 24:	1	1	3	3	7	3	7	13	22	24

We'll verify some of these values momentarily. First let's expand the other two expressions. The second expression yields seven distinct terms.

	x^6	x^5y	x^4y^2	x^3y^3	x^4yz	x^3y^2z	$x^2y^2z^2$
$(x+y+z)^6$	1	6	15	20	30	60	90
$6(x+y+z)^2(x^4+y^4+z^4)$	6·1	6·2	6·1	0	6·2	0	0
$3(x+y+z)^2(x^2+y^2+z^2)^2$	3·1	3·2	3·3	3·4	3·2	3·4	3·6
$6(x^2+y^2+z^2)^3$	6·1	0	6·3	0	0	0	6·6
$8(x^3+y^3+z^3)^2$	8·1	0	0	8·2	0	0	0
Total ÷ 24:	1	1	2	2	2	3	6

The third expression contains only four distinct terms.

	x^4	x^3y	x^2y^2	x^2yz
$(x+y+z)^4$	1	4	6	12
$6(x^4+y^4+z^4)$	6·1	0	0	0
$3(x^2+y^2+z^2)^2$	3·1	0	3·2	0
$6(x+y+z)^2(x^2+y^2+z^2)$	6·1	6·2	6·2	6·2
$8(x+y+z)(x^3+y^3+z^3)$	8·1	8·1	0	0
Total ÷ 24:	1	1	1	1

Now it's time to verify a few of these numbers. Some of them are obviously correct. For instance, if we use only one color "bead", there is only one way in which to put beads on the eight vertices, so the coefficient of x^8 in the first expression should indeed be 1. Similarly, the coefficients of x^6 and x^4 in the other two expressions should be 1. If we let exactly one of the beads be a different color, there is still only one configuration, since the vertex (or face or diagonal) that has the uniquely colored bead can be rotated into any position.

Let's try a more complicated case. Consider the term $7x^5y^2z$. Are there indeed exactly 7 ways to place one white bead, two black beads, and five "invisible" beads on the vertices of a cube? The answer is, of course, yes, and here they are. (The two black beads may be separated by an edge, or by a face diagonal, or by a space diagonal. The first two cases each have three distinct positions for the white bead; the third case has only one.)

What about, in the second expression, the term $2x^4yz$? If we paint one face white and one face black (and leave the other four unpainted), there are indeed two distinct configurations. Either the two faces touch along one edge, or they do not touch.

What can we say about the third expression? Each of the coefficients is 1; does this make sense? Notice that there are 24 ways to rotate the cube, and that each rotation permutes the diagonals in a different way. Since there are 4 diagonals, there are only 4! = 24 different ways in which they can be permuted. Therefore any permutation of the diagonals can be accomplished by some rotation in the group, and so any given combination of colors can be applied in only one way. Any other configuration using the same colors can be permuted into the first by one of the 24 rotations.

That's enough "inspection". It's time for us to move on to the "open-ended" section.

Part 'c' (extra credit, points awarded as warranted). One of the first things we should notice (though, disappointingly, few people did) is that the cycle index for the diagonals is the same as that for the symmetric group S_4, and that therefore the diagonals' permutation group must in fact be the symmetric group. (At the end of part (*b*) we described the implications of this.) We could double-check this using the formula for finding the cycle index of a symmetric group, but this would essentially duplicate page 74, so we won't bother.

Some observations we can make, but from which it is difficult to deduce any general principles, include the fact that the vertex group included two terms, generated by rotations about different types of axes, that had the same cycle form, whereas this did not occur in the other two groups. Also, each term in the vertex group yields the same coefficient for x^5y^3 as for x^6yz. As we said, it is not clear what the implications are of these similarities within the vertex group, but they are certainly worth noting.

We have already pointed out that the number of rotations remains the same regardless of the elements being permuted. We can extend this observation by noting that the number of rotations about any given type of axis, or about any single specific axis, is invariant among the three groups. This may appear to be a case of stating the obvious, but it is an important property nevertheless.

One person noted that the fact that the midpoints of the faces of a cube form the vertices of an octahedron was merely a particular instance of a general useful phenomenon. When two sets correspond to one another in this fashion, they will necessarily have the same cycle index under any given form of permutation. For instance, were we to work out the cycle index for the permutations of the faces of a cube under both rotation and reflection, we would simultaneously produce the cycle index for the vertices of an octahedron under rotation and reflection.

One very significant observation is the correlation between the degree of rotation (expressed in radians) and the corresponding term of the cycle index. In every instance (at least, every instance in this particular problem), a rotation of $2\pi/k$ radians yields a permutation whose largest cycle is of order k. In fact, the permutation consists of one or more cycles of order k and perhaps some cycles of order 1, and no others. It is clear why this should be the case. A cycle of order k indicates an operation that, if performed k times, restores the original state. The operation of rotating an object through an angle of $2\pi/k$ radians, if performed k times, does indeed restore the object to its original position. Furthermore, repeating the operation fewer than k times will *not* restore the original position, so there should be no cycles of order $k-1$ or smaller, except that a cycle can be of order 1, indicating an element that is not affected at all by this particular rotation.

Having explained why a rotation through $2\pi/k$ radians should always yield at least one cycle of order k and no cycles of orders other than 1 and k, we should point out that this is not strictly true. For one thing, it's true only of physical rotations of physical objects whose shapes are not subject to change. That is, if we rotate one part of the object through $180°$, there mustn't be some other part that changes position by only 120°. For another, there are times when a rotation of less than 360° may restore the original position. For instance, if we were to consider the <u>diagonals</u> of a rotating hexagon (considering only the three diagonals that join opposing pairs of vertices), we would discover that a 180° rotation causes all three diagonals to land back in their original positions. Still, this sort of thing is probably a pathological case. In most cases the observation in the preceding paragraph is a valid and useful one.

One person pointed out a useful special case of the general figure inventory. Specifically, let the figure inventory be

$$f(x) = \sum_{r=0}^{\infty} a_r x^r = 3x.$$

Since there is only one variable, it means we are dealing with a single "property", such as "any color". The coefficient, which could be any positive integer constant, is the number of variations on the property, such as the number of different colors. Substituting this figure inventory into the cycle index, we find that all the terms involve x^d, where d is the degree of the group. The coefficient of x^d is the number of distinct configurations in which each element being permuted has been assigned one of the variations on the property. *E.g.*, substituting this figure inventory into the cycle index for the faces of the cube would result in the single term $57x^6$, indicating that there are 57 distinct ways of painting the faces of a cube, where each face can be painted any of 3 available colors.

One brave soul attempted to present a rough idea of *why* substituting the figure inventory into the cycle index works the way it does. He did a reasonable job, considering the difficulty of the undertaking. Since he was moved to try explaining the theory behind all this, can we do less? Probably. But let's see what we can do. We won't attempt to prove anything here, but we'll try to present some insights on what's happening when the figure inventory is substituted into the cycle index, and why it might be reasonable to expect it to yield the results it does. If this explanation is too confusing, then skip it, but we feel it's worth the try.

Consider the cycle index for rotations of the vertices of a regular hexagon. As we saw in chapter 6, this cycle index is

$$\frac{f_1^6 + 2f_6 + 2f_3^2 + f_2^3 + 3f_1^2f_2^2 + 3f_2^3}{12}.$$

We'll use the figure inventory $x + y$, representing two types of object. We'll denote the two types in the diagrams that follow by circling vertices of one type and leaving the others empty. Let's suppose we want to find the number of configurations in which exactly two vertices are of type x (circled). We know there are three such configurations.

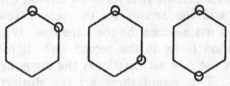

We don't have any systematic formula that counts just these three configurations, so instead we'll count something similar. We'll count each configuration, rotated every possible way. For example, the center configuration above can be rotated according to each of the 12 permutations in the group, yielding the 12 configurations shown below. The top left configuration comes from the identity permutation, the other five in the top row come from rotations around the center, the leftmost three in the second row result from rotations about axes drawn through two opposing vertices, and the remaining three result from axes drawn through two opposing edges.

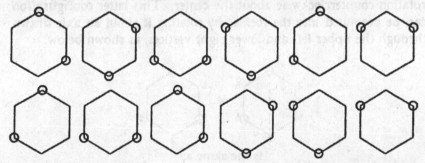

As you can see, not all of these configurations "look" different. That is, different rotations can yield the same configuration. (The actual rotations are not identical, however. It is only this particular configuration that is indistinguishable under these rotations.) We wish to count all 12 of these configurations, even though some of them look identical to others. If we can count all 12 of them and likewise count all 12 rotations of each of the other configurations for two circles, we can then divide by the total number of rotations and we will have the answer we wanted. So let's see how we can go about counting these 12 configurations.

We start by separating the 12 configurations into two sets. The first set contains as many configurations as possible such that no two of them "look" alike, and the second set contains the rest. In this particular example, the first set could contain the top row of six configurations, and the second set the second row. In general, however, the two sets need not be the same size. What does it mean for a configuration to be in the second set? It means there is a configuration in the first set that looks the same, *i.e.*, has the same vertices circled. This means there are two <u>distinct</u> rotations that permute the "original" orientation (upper left corner in the preceding diagram) into this form. (We can see that the two rotations are indeed distinct, and hence the resulting look-alike configurations are in fact different, by numbering all six vertices instead of merely circling two of them.) Since the two rotations are distinct, there must exist some rotation, *not* the identity permutation, that maps the configuration in the first set into that in the second set. For instance, the bottom left configuration of the twelve, which resulted from a rotation about an axis drawn through the top and bottom vertices, "looks" the same as one in the top row, which resulted from a 120° rotation counterclockwise about the center. This latter configuration can be permuted into the former by rotating it about an axis drawn through the upper left and lower right vertices, as shown below.

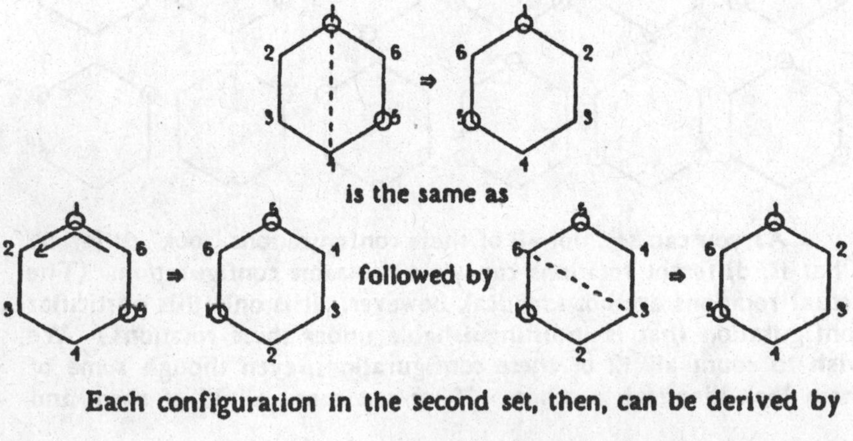

is the same as

followed by

Each configuration in the second set, then, can be derived by taking a configuration in the first set and applying a rotation that doesn't change the "appearance" of the configuration. When we perform a rotation that interchanges the two circles, the appearance remains unchanged even though we have generated a different configuration. Furthermore, if we have any rotation that doesn't change the appearance of some configuration, and the configuration is in the first set, we can apply the rotation to that configuration and come up with another configuration, which must be in the second set. Therefore we have established a one-to-one correspondence between the configurations in the second set and rotations that do not change the appearance of some configuration.

Observe that, if we consider all distinct configurations (in this case there are 3 of them) and look at all their rotations, the resulting set of configurations must include all possible combinations of two vertices selected from the six available. Hence if we combine all the sets of configurations no two of which look alike, we must end up with all possible combinations, once each. We know how many such combinations there are; in our particular example it's $\binom{6}{2} = 15$. All we have to do now is count how many configurations are in the "second sets"; in other words, how many ways are there to take a configuration and rotate it in such a way that it appears unchanged? This is where the cycle index comes in.

Consider, for example, the term $f_1{}^2 f_2{}^2$, which results from a rotation about an axis that goes through two opposing vertices, as shown here for one particular case.

Replacing the exponents by repeated multiplication, then substituting the figure inventory, we get

$$(x+y) \cdot (x+y) \cdot (x^2+y^2) \cdot (x^2+y^2)$$

The coefficient of $x^2 y^4$ in this product is equal to the number of ways of selecting one element from each of the 4 factors such that the product of the selected elements is x^2y^4. For instance, we could select the y from the first factor, y from the second, x^2 from the third, and y^2 from the fourth. But let's remember what these factors represent. A factor of the form (x^k+y^k) indicates that the rotation yielding this particular term contains a cycle of order k. When we select the y^k out of such a factor, it corresponds to a configuration in which all k elements of that cycle are of type y. Thus, in our example, the selection of $y \cdot y \cdot x^2 \cdot y^2$ corresponds to a configuration in which the two cycles of order 1 both consist of "empty" vertices, as does the second of the two cycles of order 2, while the first cycle of order 2 consists of two "circled" vertices. This configuration is diagrammed below (left). Notice that this rotation doesn't change the appearance of this configuration. The other terms contributing to the coefficient of x^2y^4 are $y \cdot y \cdot y^2 \cdot x^2$ and $x \cdot x \cdot y^2 \cdot y^2$, which correspond to the other two configurations shown below (center and right).

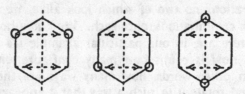

Let's summarise what this means: the coefficient of x^2y^4 in any given term of the cycle index is equal to the number of ways of placing two x's and four y's such that each cycle in the permutation represented by that term contains only x's or only y's. That is, if we count the number of ways of placing two x's and four y's such that no cycle in the particular term contains elements of both types, we get the coefficient of x^2y^4. Finally, we observe that it is exactly these situations that we wish to count, since a permutation will change the appearance of a configuration if and only if one or more of its cycles involves elements of two different types. (Such a cycle must of necessity permute at least one element of type x into a position formerly occupied by an element of type y, thereby changing the appearance of the configuration.)

Now consider the entire cycle index. The identity term, after substituting the figure inventory, is $(x+y)^6$. The coefficient of x^2y^4 in this term is $\binom{6}{2}$, the number of different-looking configurations. In each of the remaining terms of the cycle index, by the reasoning outlined above, the coefficient of x^2y^4 is equal to the number of ways in which the rotation corresponding to that particular term can transform one of these $\binom{6}{2}$ configurations into another configuration that looks the same but is really a different permutation of the six vertices. When we sum over all the terms in the cycle index, the coefficient of x^2y^4 thus comes out to be the number of <u>distinct</u> configurations (in our example, 3), multiplied by the number of rotations that can be applied to each of these configurations (in our example, 12). We divide by the number of rotations and there's our answer.

We hope that the preceding discussion provides some faint illumination as to the inner workings of Pólya's Theory of Counting. This is really the sort of thing that should be discussed in much greater detail, with much greater formality, and in the interactive setting of a classroom. Lacking the time for this, however, we hope this has given you at least some insight into the theory underlying the method.

Problem 2 (20 points total).

Part 'a' (20 points). The identity to be proved in this problem is actually true only when both m and n are non-negative integers. Fortunately, everyone seemed to realise this, so our failure to mention it on the midterm handout didn't lead to any difficulties. Three different forms of "proof" showed up among the papers handed in. One of these methods is not quite valid, and a few points were deducted for using it. We'll present it here in order to be able to point out what's wrong with it.

It typically ran something like this. We know, by the recursive definition of the binomial coefficients, that

$$\binom{n+1}{r} = \binom{n}{r-1} + \binom{n}{r}$$

for any integers n and r such that $0 < r < n+1$. This equation continues to hold even if $r \geq n+1$ (assuming $n+1>0$), if we assume $\binom{s}{t} = 0$ whenever s and t are integers and $0 \leq s < t$. This is a reasonable assumption to make, since after all there are zero ways to choose a subset of t elements from a set with fewer than t elements. By repeated application of this formula, we deduce that

$$\binom{n+1}{m+1} = \binom{n}{m} + \binom{n}{m+1}$$

$$= \binom{n}{m} + \binom{n-1}{m} + \binom{n-1}{m+1}$$

$$= \binom{n}{m} + \binom{n-1}{m} + \binom{n-2}{m} + \binom{n-2}{m+1}$$

$$\vdots$$

$$= \binom{n}{m} + \binom{n-1}{m} + \binom{n-2}{m} + \cdots + \binom{0}{m} + \binom{0}{m+1}.$$

Assuming $m \geq 0$, $\binom{0}{m+1}$ must be zero, so we find that

$$\binom{n+1}{m+1} = \binom{n}{m} + \binom{n-1}{m} + \binom{n-2}{m} + \cdots + \binom{0}{m} = \sum_{k=0}^{n} \binom{k}{m},$$

which is the desired identity. The problem with this "proof" is that it's not really a formal proof, since it involves an indefinite number of steps. This reasoning would suffice in an informal discussion, or perhaps as an answer to a request to "show" instead of "prove", but in a proof you should be a bit more cautious. One reason this was emphasised so much in grading the midterm is that, if you're not careful, such an approach can result in "proving" things that are in fact false! The problem can arise in two ways. You may fail to observe that one of the steps glossed over in the " . . . " is a special case for which the general equation doesn't apply, or you might attempt to use the " . . . " to represent an infinite sequence of steps. Mathematical induction, which is the formal representation for this sort of proof, proves things only for any <u>finite</u> number of repetitions of the inductive step. It is not difficult to find statements that are true for any finite n, and can be proved by induction, but are false if n is infinite.

Anyway, that's enough about that. Here's the formal proof using induction. We first consider the case $n = 0$. In this case,

$$\binom{n+1}{m+1} = \binom{1}{m+1} = \binom{0}{m} + \binom{0}{m+1} = \binom{0}{m} = \sum_{k=0}^{0} \binom{k}{m}$$

(since $\binom{0}{m+1} = 0$). So the formula is true for all m when $n = 0$. Now we assume that it's true for all non-negative integers $n \leq s$ for some value s. We need to show that the formula is then true for $n = s+1$.

$$\binom{n+1}{m+1} = \binom{s+2}{m+1} = \binom{s+1}{m} + \binom{s+1}{m+1} \quad \text{[by the recursion formula]}$$

$$= \binom{s+1}{m} + \sum_{k=0}^{s} \binom{k}{m} \quad \text{[by the induction hypothesis]}$$

$$= \sum_{k=0}^{s+1} \binom{k}{m}$$

By induction, then, the formula is true for any finite integer $n \geq 0$.

One never knows what interesting things will turn up on an examination paper. Two people found a completely different way of approaching this problem. Their method is quite reasonable and valid, and it is interesting enough to warrant our reproducing it here. It uses generating functions instead of induction. (These are, after all, the two most common ways of dealing with an arbitrarily large sequence of values.) We introduce the following generating function involving m as the exponent:

$$g(x) = \sum_{m=0}^{\infty} \left(\sum_{k=0}^{n} \binom{k}{m} \right) x^{m}.$$

We first interchange the order of the two summations.

$$g(x) = \sum_{k=0}^{n} \sum_{m=0}^{\infty} \binom{k}{m} x^{m}$$

The terms for $m > k$ are zero, so we can apply the binomial theorem and find

$$g(x) = \sum_{k=0}^{n} (1+x)^{k},$$

which is simply a geometric series. We know what the sum of a geometric series is—we encountered it in chapter 3 (page 13). So

$$g(x) = \frac{1 - (1+x)^{n+1}}{1 - (1+x)}$$

$$= \frac{(1+x)^{n+1} - 1}{x}$$

$$= x^{-1} \left[\left(\sum_{k=0}^{n+1} \binom{n+1}{k} x^{k} \right) - 1 \right].$$

In this new summation, the term for $k = 0$ is $\binom{n+1}{0}x^0 = 1$, so we can cancel this against the "-1" that follows the summation.

$$g(x) = x^{-1}\left(\sum_{k=1}^{n+1} \binom{n+1}{k} x^k\right)$$

$$= \sum_{k=1}^{n+1} \binom{n+1}{k} x^{k-1}$$

$$= \sum_{k=0}^{n} \binom{n+1}{k+1} x^k$$

By equating this with our original definition of $g(x)$ and equating the coefficients of x^m in the two forms, we produce the desired identity.

Part 'b' (extra credit, points awarded as warranted). Since we're supposed to apply the identity from part (a), we must somehow transform the summation of k^t into a summation involving binomial coefficients. This is easily done using one of the formulas derived in chapter 5:

$$x^n = S_1^n x + S_2^n x(x-1) + S_3^n x(x-1)(x-2) + \cdots + S_n^n x(x-1)(x-2) \ldots (x-n+1).$$

(Actually, this formula is not quite correct, and we might as well take this opportunity to correct it. The right-hand side should include the term S_0^n; otherwise the formula is invalid for $n = 0$. Since $S_0^0 = 1$ and $S_0^n = 0$ for $n > 0$, this additional term takes care of the case $n = 0$ without affecting the other cases. Similarly, the other boxed formula in that chapter should include the term $(-1)^n \mathcal{S}_0^n$.)

Using this formula, we find

$$\sum_{k=0}^{n} k^t = \sum_{k=0}^{n} \sum_{m=0}^{t} S_m^t \binom{k}{m} m!$$

$$= \sum_{m=0}^{t} \left(S_m^t m! \sum_{k=0}^{n} \binom{k}{m}\right)$$

$$= \sum_{m=0}^{t} S_m^t m! \binom{n+1}{m+1}.$$

Since $\binom{n+1}{m+1}$ is essentially a polynomial in n, this would be sufficient as an answer to the problem. (It would be a good idea to check the

answer for a few small values of t, of course.) A few people went
further by applying the "other" formula from the end of chapter 5:

$$\delta_n^n x^n - \delta_{n-1}^n x^{n-1} + \delta_{n-2}^n x^{n-2} - \cdots + (-1)^{n-1} \delta_1^n x + (-1)^n \delta_0^n = \binom{x}{n} n!.$$

We have to start by backtracking a bit, lest we end up with a
polynomial in terms of $(n+1)$ instead of n. We observe that

$$\sum_{k=0}^{n} k^t = n^t + \sum_{k=0}^{n-1} k^t$$

$$= n^t + \sum_{m=0}^{\infty} S_m^t \, m! \, \binom{n}{m+1}$$

$$= n^t + \sum_{m=0}^{\infty} S_m^t \, \binom{n}{m+1} \, \frac{(m+1)!}{m+1}$$

$$= n^t + \sum_{m=0}^{\infty} S_m^t (m+1)^{-1} \cdot \left(\sum_{r=0}^{m+1} (-1)^{m+1-r} \delta_r^{m+1} n^r \right).$$

Note that we have extended the summation on m so that it
runs to infinity. This is valid since S_m^t is zero for $m > t$. There are
various ways we could go from here. One of the nicer ways (nicer
since it eliminates that annoying n^t term) starts by applying the
recursion formula for Stirling numbers of the second kind,

$$S_k^{n+1} = S_{k-1}^n + k S_k^n.$$

From this we find that $S_m^t = S_{m+1}^{t+1} - (m+1) S_{m+1}^t$, and thus

$$\sum_{k=0}^{n} k^t = n^t + \sum_{m=0}^{\infty} S_{m+1}^{t+1} (m+1)^{-1} \cdot \left(\sum_{r=0}^{m+1} (-1)^{m+1-r} \delta_r^{m+1} n^r \right)$$
$$- \sum_{m=0}^{\infty} S_{m+1}^t \cdot \left(\sum_{r=0}^{m+1} (-1)^{m+1-r} \delta_r^{m+1} n^r \right).$$

Since

$$\sum_{m=0}^{\infty} S_{m+1}^t \cdot \left(\sum_{r=0}^{m+1} (-1)^{m+1-r} \delta_r^{m+1} n^r \right)$$

$$= \sum_{m=0}^{\infty} S_{m+1}^t \, \binom{n}{m+1} \, (m+1)!$$

$$= \sum_{m=1}^{\infty} S_m^t \binom{n}{m} m!$$

$$= \Big(\sum_{m=0}^{\infty} S_m^t \binom{n}{m} m! \Big) - S_0^t \binom{n}{0} 0!$$

$$= n^t - S_0^t.$$

we can deduce that

$$\sum_{k=0}^{n} k^t = S_0^t + \sum_{m=0}^{\infty} S_{m+1}^{t+1} (m+1)^{-1} \cdot \Big(\sum_{r=0}^{m+1} (-1)^{m+1-r} \delta_r^{m+1} n^r \Big)$$

$$= S_0^t + \sum_{m=1}^{\infty} S_m^{t+1} m^{-1} \cdot \Big(\sum_{r=0}^{m} (-1)^{m-r} \delta_r^m n^r \Big).$$

Since δ_0^m is zero for all $m \geq 1$, we can start the r-summation at 1 instead of 0. The two summations combined are then a summation over all integers m and r such that $1 \leq r \leq m$. We therefore sum the same terms if we let r range from 1 to infinity and let m range from r to infinity.

$$\sum_{k=0}^{n} k^t = S_0^t + \sum_{r=1}^{\infty} \sum_{m=r}^{\infty} S_m^{t+1} \frac{(-1)^{m-r}}{m} \delta_r^m n^r$$

Knowing that $S_m^{t+1} = 0$ if $m > t+1$, we can stop the second summation at $m = t+1$. Having done this, we find that the summation becomes null if $r > t+1$, so we can stop the first summation at $r = t+1$.

$$\sum_{k=0}^{n} k^t = S_0^t + \sum_{r=1}^{t+1} \Big[n^r \cdot \sum_{m=r}^{t+1} S_m^{t+1} \frac{(-1)^{m-r}}{m} \delta_r^m \Big]$$

We finally have a single polynomial in n. The coefficients are not particularly simple, but we can quickly work them out for any given value of t. Let's check a few cases. If $t = 0$ we get

$$\sum_{k=0}^{n} k^0 = \sum_{k=0}^{n} 1 = n+1.$$

The formula says that the sum should be

$$S_0^0 + n^1 S_1^1 \frac{(-1)^{1-1}}{1} \delta_1^1.$$

which does indeed equal $n+1$. For $t = 1$ we know that

$$\sum_{k=0}^{n} k^1 = \sum_{k=0}^{n} \binom{k}{1} = \binom{n+1}{2}$$

by using the formula from part (a). We expect this to equal

$$S_0^1 + n(S_1^2 \delta_1^1 - \tfrac{1}{2}S_2^2 \delta_1^2) + n^2(\tfrac{1}{2}S_2^2 \delta_2^2)$$

$$= 0 + \tfrac{1}{2}n + \tfrac{1}{2}n^2,$$

which again checks. The formulas for $t = 2$ and $t = 3$ work out to be

$$\sum_{k=0}^{n} k^2 = 0 + \frac{n}{6} + \frac{n^2}{2} + \frac{n^3}{3} = \frac{n(n+1)(2n+1)}{6}$$

and $\sum_{k=0}^{n} k^3 = 0 + 0 \cdot n + \frac{n^2}{4} + \frac{n^3}{2} + \frac{n^4}{4} = \left(\sum_{k=0}^{n} k\right)^2,$

which also turn out to be correct.

One person even computed some of the individual coefficients. The coefficient of n^{t+1} (which is the highest power of n for which the coefficient is non-zero) is always $1 / (t+1)$. More surprisingly, for $t > 0$, the coefficient of n^t is always $\tfrac{1}{2}$, independent of t. This person then went on to comment, "It turns out that following coefficients can be expressed in terms of Bernoulli numbers, but this is probably beyond the scope of the midterm." She was quite right—on both counts.

Problem 3 (30 points total).

Let us denote the n digits in any particular n-digit number by the variables $d_1, d_2, \ldots, d_n$. How many such numbers do not start with a zero and have no adjacent pairs of digits alike? Let us find the answer first without using PIE.

The first digit (d_1) can be any digit other than a zero. Hence it has 9 possible values. The second digit, d_2, can be any digit different from d_1. Thus, for each possible value of d_1, d_2 also has 9 possible values. Similarly, for any digit d_i there are 9 possible values, regardless of the values of the other digits. (The 9 possible values depend on the values of the other digits, but there are always 9 of them.) By the product rule from way back in chapter 2, then, the total number of combinations must be

$$\prod_{i=1}^{n} 9,$$

which is simply 9^n. That was easy; now for the hard part! (Even though this problem is easier to solve without PIE than with it, we felt it was worth including on the midterm. It's not particularly complicated even using PIE, and it lets you check your result using the non-PIE approach.)

Most people did fairly well applying PIE to this problem. Those who didn't seemed confused as to exactly how PIE works. Several people attempted to define N_α, N_β, etc., without specifying exactly what α, β, etc., were. One particularly popular mistake was defining N_α to be the number of n-digit numbers with at least one pair of adjacent digits alike, N_β the number of numbers with some three adjacent digits alike, and so forth. This approach quickly runs into trouble when it comes time to define $N_{\alpha\beta}$. People who somehow managed to hedge past this found that numbers such as "3344", containing two different pairs of matching digits, were counted twice by N_α but not at all by N_β, N_γ, etc., resulting in an answer that was too small for $n \geq 4$.

Here are some rules of thumb to keep in mind any time you intend to use PIE:

[1]   Start by establishing *exactly* what you intend to have as the properties α, β, etc.

[2]   The *union* of these properties (that is, the union of the sets of elements having each property) *must* consist of precisely those elements that you do *not* wish to count.

[3]   Try to choose properties that are "interchangeable". That is, see if you can arrange things so that $N_\alpha = N_\beta = \cdots = N_\lambda$, and $N_{\alpha\beta} = N_{\alpha\gamma} = N_{\beta\gamma} = \cdots$, and so forth. This sort of symmetry will make the PIE formula much simpler to evaluate.

[4]   Choose α, β, and the other properties such that the individual
terms (N_α, N_β, . . . , $N_{\alpha\beta}$, etc.) are easy to evaluate.

For this particular problem, we will let α_1 be the property that
d_2 is the same as d_1, α_2 the property that d_3 is the same as d_2, . . . ,
and in general α_i the property that d_{i+1} is the same as d_i. There are
a total of $n-1$ such properties. This clearly follows rules [1] and [2];
we'll soon see that [3] and [4] are also met.

Consider any combination of k properties: α_{i_1}, α_{i_2}, . . . , α_{i_k}.
How many n-digit numbers have this particular set of properties?
That is, what is $N(\alpha_{i_1},\alpha_{i_2},...,\alpha_{i_k})$? (Pardon our not using subscripts
on the N, but subscripts on subscripts on subscripts are absolutely
illegible!) The first digit, d_1, can still be any digit other than zero.
Each of the remaining $n-1$ digits can be any of the ten decimal
digits, *except* that exactly k of them are constrained to be the same as
their immediately preceding digits. Thus there are $9 \cdot 10^{n-1-k}$ such
n-digit numbers, and $N(\alpha_{i_1},\alpha_{i_2},...,\alpha_{i_k}) = 9 \cdot 10^{n-1-k}$. Furthermore, since
there are $n-1$ properties, there are $\binom{n-1}{k}$ such terms. This formula is
valid even when $k = 0$; that is, there are a total of $9 \cdot 10^{n-1}$ n-digit
numbers.

Applying the PIE formula, we find

$$N_0 = 9 \cdot 10^{n-1} - (n-1) \cdot 9 \cdot 10^{n-2} + \binom{n-1}{2} \cdot 9 \cdot 10^{n-3} - \cdots \pm \binom{n-1}{n-1} \cdot 9 \cdot 10^0$$

$$= 9 \cdot \sum_{k=0}^{n-1} \binom{n-1}{k} (-1)^k \, 10^{n-1-k}$$

$$= 9 \cdot (10-1)^{n-1}$$

by the binomial theorem. So $N_0 = 9^n$, as expected.

February 23. This was Tarjan's first lecture, and he started by announcing what he intended to cover during his seven lectures. Specifically, he said he would be discussing miscellaneous problems in existential and constructive combinatorics, with the emphasis on the constructive side. This chapter, on the other hand, deals almost entirely with existential combinatorics, although some of the proofs are constructive.

We encountered Ramsey Theory briefly in chapter 7, where we looked at one of its simplest cases. If there are six people at a party, either there are three people who know each other or there are three none of whom know each other. Tarjan went over the proof of this result. We won't take the time to include it here; you're encouraged to refer to pages pages 87 through 89 if you wish to review that discussion. As before, we'll use solid and dotted lines to represent two different "colors" in these notes.

One of the things we'd like to do is generalise this result to the case where there are n people who all know each other or n none of whom know each other, for some arbitrary value of n. We'll get to this in a moment, but first we want to look at a different question. Is there another graph with the 3-people property? That is, is there some other graph besides K_6 that, if its edges are colored using two colors, must contain a monochromatic triangle? Clearly, any graph that "contains" a K_6 (*i.e.*, has six vertices each of which has an edge leading to each of the other five) will contain a monochromatic triangle, since those six vertices must include one. Are there any graphs that do not contain K_6 but must nevertheless include a monochromatic triangle?

In fact there are; one such graph is shown at the top of page 117. It is created by taking a triangle and a pentagon and adding edges connecting every vertex in the triangle to every vertex in the pentagon.

Tarjan left it an exercise (as distinguished from a homework assignment) to prove that this graph contains no K_6. This is easy enough to do: any six vertices must include at least three from the

G. Pólya et al., *Notes on Introductory Combinatorics*, Modern Birkhäuser Classics,
DOI 10.1007/978-0-8176-4953-1_9, © Birkhäuser Boston, a part of Springer
Science+Business Media, LLC 2010

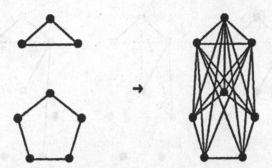

pentagon, and some two of those three must lack a mutual edge. To prove that this graph, if two-colored, must include a monochromatic triangle, we start by looking at the three vertices that make up the upper triangle. If the edges connecting these vertices are of a single color, we've found a monochromatic triangle. Otherwise, there must be one edge of one color and two of the other color. Since the three edges are symmetric with respect to the rest of the graph, and the two colors are interchangeable, let's assume the graph looks like this (edges not shown haven't been assigned a color yet):

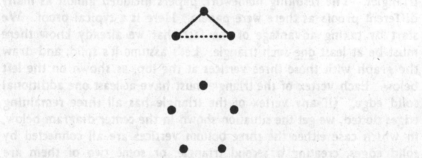

Consider the two vertices joined by the dotted edge. If either of these vertices has two solid edges leading to adjacent vertices of the pentagon, we get the situation shown on the left at the top of page 118. In this case, the two vertices in the pentagon must be joined by a dotted edge, lest we get a solid triangle. They must also be joined to the topmost vertex of the triangle by dotted edges, for the same reason. But this would give us a dotted triangle, as shown in the center diagram. On the other hand, we <u>can</u> have two solid edges from a vertex in the triangle to <u>non</u>-adjacent vertices in the pentagon (rightmost diagram). From either of the vertices on the dotted edge in the upper triangle, there can be no more than two solid edges leading to vertices in the pentagon. (If there were three

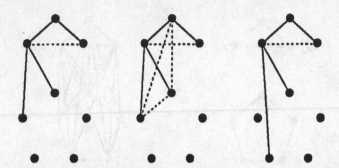

solid edges they would necessarily include two that lead to adjacent vertices.) Thus each of these two vertices in the triangle has at least three dotted edges leading to the pentagon. This makes a total of six dotted edges, so at least two of them must go to the same vertex in the pentagon. These two edges, together with the dotted edge in the upper triangle, form a monochromatic triangle.

As homework, Tarjan asked for a proof that if the edges of K_6 are two-colored there must result at least <u>two</u> monochromatic triangles. The resulting homework papers included almost as many different proofs as there were papers. Here is a typical proof. We start by taking advantage of the fact that we already know there must be at least <u>one</u> such triangle. Let's assume it's solid, and draw the graph with those three vertices at the top, as shown on the left below. Each vertex of the triangle must have at least one additional solid edge. (If any vertex of the triangle has all three remaining edges dotted, we get the situation shown in the center diagram below, in which case either the three bottom vertices are all connected by solid edges, creating a second triangle, or some two of them are connected by a dotted edge, creating a dotted triangle.) If any vertex in the first triangle has <u>more</u> than one additional solid edge, we get the situation shown on the right below, in which we either have a

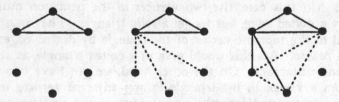

dotted triangle as shown, or one of the edges shown as dotted must instead be solid, giving a solid triangle. Thus, for each of the points of the first triangle, the three remaining edges must include *exactly* one solid edge and two dotted edges. In addition, each of the three additional solid edges (one from each vertex of the triangle) must go to a <u>different</u> vertex, lest we get another solid triangle (below left). Since the three lower vertices are indistinguishable at this point, we get the situation shown in the center diagram below. If we are to avoid having another solid triangle, certain edges must be dotted, as shown on the right below. But then, if we are to avoid getting a dotted triangle, each edge connecting two of the three lower vertices must be solid. This gives us a second solid monochromatic triangle. This construction also shows that it is possible to two-color a K_6 so as to have exactly two monochromatic triangles. It is not known, for general n, what the minimum number of monochromatic triangles is in a two-colored K_n.

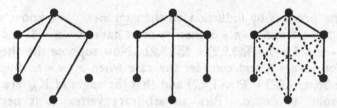

Getting back to the case of proving the existence of a single monochromatic triangle in K_6, how can we generalise this result and its proof? How large must a graph be so that when its edges are two-colored there must be a complete monochromatic subgraph of k vertices? (A triangle, of course, is a complete graph of 3 vertices.) This time it was Tarjan's turn to come up with a problem-solving aphorism: "*Sometimes to get a result you have to ask a more general question.*" In order to prove that a sufficiently large graph always exists, we must generalise the problem still more.

We define $\Re(m,n,2)$ to be the <u>minimum</u> integer such that, if $N \geq \Re(m,n,2)$ and each edge of K_N is "colored" solid or dotted, then there is either a solid K_m or a dotted K_n. (The "2" denotes that we're coloring edges, which correspond to pairs of vertices. This will be important later.) It is assumed that m and n are positive integers. We saw in chapter 7 that it is possible to two-color a K_5 graph without getting any monochromatic triangles, and that a two-colored K_6 always has such a triangle, so we know that $\Re(3,3,2) = 6$. What other values of $\Re$ can we readily ascertain? Consider $\Re(m,2,2)$. If we have a K_{m-1} graph, we can color all its edges solid, and we will have neither a solid K_m nor a dotted K_2. If we two-color the edges of K_m, however, then either we have a dotted edge (which would be a dotted K_2, since K_2 is a single edge) or else all the edges are solid, giving us a solid K_m. Hence $\Re(m,2,2) = m$. Similarly, $\Re(2,n,2) = n$. (In general, $\Re(m,n,2) = \Re(n,m,2)$, since the definition is symmetric with respect to m and n.)

We now come to Ramsey's Theorem, version 1: For all m and $n \geq 2$, $\Re(m,n,2)$ exists and satisfies the relation

$$\Re(m,n,2) \leq \Re(m,n-1,2) + \Re(m-1,n,2)$$

for all m and $n \geq 3$.

The proof is by induction on the sum $m+n$. We know, when m and n are ≥ 3 and $m+n = 6$, that we must have $m = n = 3$, and that $\Re(3,3,2) = 6 \leq 3 + 3 = \Re(3,2,2) + \Re(2,3,2)$. Now suppose the theorem is true for $m+n < t$, and consider the case when $m+n = t$. Suppose that $N \geq \Re(m,n-1,2) + \Re(m-1,n,2)$ and that the edges of K_N are each colored solid or dotted. Pick an arbitrary vertex; it is perforce connected to every other vertex in the graph, and there are at least $\Re(m,n-1,2) + \Re(m-1,n,2) - 1$ of them. Of the edges leading to these vertices, either at least $\Re(m,n-1,2)$ of them are dotted, or at least $\Re(m-1,n,2)$ of them are solid. Why is this? Well, if it weren't so, then there would be at most $\Re(m,n-1,2) - 1$ dotted edges, and at most $\Re(m-1,n,2) - 1$ solid edges. But this would give a total of at most $\Re(m,n-1,2) + \Re(m-1,n,2) - 2$ edges, and we've said there must be at least one more than that. So suppose $\Re(m-1,n,2)$ of the edges are solid. Consider the vertices at the "other ends" of these edges. (Remember that all these edges are coming from a single vertex.) According to our induction hypothesis, these $\Re(m-1,n,2)$ vertices and their interconnecting edges must include either a solid K_{m-1} or a dotted K_n. If the latter, we're done, since the original K_N contains this same K_n. If there's a solid K_{m-1}, then the original K_N contains a solid K_m consisting of these $m-1$ vertices together with the vertex we selected at the start, since that vertex is known to be connected by solid edges to each of the $m-1$ vertices. The case when the vertex we start with has $\Re(m,n-1,2)$ dotted edges is handled similarly, and is left as an exercise.

You should be sure that you understand this proof before proceeding; the same approach will be used later on to prove the more general versions of Ramsey's Theorem. Notice how this proof generalises the method that we used to determine $\Re(3,3,2)$. Note also that the proof forces us to consider $\Re(m,n,2)$ for $m \neq n$.

Using this theorem, we can compute upper bounds on the values of the Ramsey numbers $\Re(m,n,2)$. Here's a table showing these bounds for the first few values of m and n.

m \ n	2	3	4	5	6	7
2	2	3	4	5	6	7
3	3	6	10	15	21	28
4	4	10	20	35		
5	5	15	35			

These values should look familiar—they are simply the binomial coefficients. This isn't terribly surprising, since the upper bound given by the theorem is the same as the recursive definition of the binomial coefficients. Ramsey's Theorem (version 1) tells us that $\Re(m,n,2) \leq \binom{m+n-2}{m-1}$. On the other hand, certain Ramsey numbers (not many) are known precisely. The following table, which is copied from [Harary], shows all the values currently known (not counting cases where m or $n = 2$, and omitting some values which are identical to others by symmetry).

$\Re(m,n,2)$

m \ n	2	3	4	5	6	7
2	2	3	4	5	6	7
3	3	6	9	14	18	23
4	4	9	18			
5	5	14				

Notice that the bound given by the theorem gets progressively worse. There is no general method known for finding these (and hence additional) numbers. Computing a few more entries for the above table is usually sufficient for a thesis; finding a general method would bring instant fame (at least within the world of mathematics).

How can we generalise this idea still further? We've pointed out that, in coloring the edges, we're actually assigning a color to each pair of vertices (where two pairs sharing a common vertex does *not* imply that the two pairs must be assigned the same color; it is the *pairings* that are being assigned colors, *not* the individual vertices). Suppose we instead assign a color to every possible subset consisting of r vertices? We shall call such a set an r-subset.

We define $\Re(m,n,r)$ to be the minimum integer such that, if each r-subset of a set containing $N \geq \Re(m,n,r)$ elements is colored either solid or dotted, then either there is an m-subset all of whose r-subsets are solid, or there is an n-subset all of whose r-subsets are dotted. As an example, let r be 3, and suppose we have a set of five elements, $\{a,b,c,d,e\}$. There are $\binom{5}{3} = 10$ 3-subsets. If we let the six subsets $\{a,b,c\}$, $\{a,b,d\}$, $\{a,c,d\}$, $\{a,c,e\}$, $\{a,d,e\}$, and $\{c,d,e\}$ be "solid", and the remaining four 3-subsets "dotted", then there is a 4-subset, $\{a,c,d,e\}$, such that all 3-subsets of this 4-subset are solid. On the other hand, it is certainly possible to assign "colors" to the 3-subsets in such a way that there is no such 4-subset, so $\Re(4,4,3)$ is not 5. (In fact, no one knows what the value of $\Re(4,4,3)$ is!)

Let's note some special cases for which the values of $\Re(m,n,r)$ are known. For instance, what about $\Re(m,r,r)$? We claim that $\Re(m,r,r) = m$. (It is obvious that $\Re(m,r,r)$ cannot be less than m.) If we have a set of m (or more) elements, then either there is at least one dotted r-subset, in which case those r elements form an r-subset "all of whose r-subsets are dotted" (an r-subset has exactly one r-subset, namely itself), or else all of the r-subsets are solid, in which case we have m elements all of whose r-subsets are solid. Similarly, by symmetry, $\Re(r,n,r) = n$. Suppose $r = 1$? In this case we are simply coloring each vertex individually. We claim $\Re(m,n,1) = m+n-1$ for all m and $n \geq 1$. If a set has this many elements, and each is colored either solid or dotted, then we must have either m solid or n dotted elements. If we didn't, then we could have at most $m-1$ solid and $n-1$ dotted elements, which accounts for only $m+n-2$ of them. This special case of Ramsey's Theorem (which we're about to state in its general form) is called the <u>pigeonhole principle</u>: if we distribute $m+n-1$ objects into two categories, either the first category contains m or more objects or else the second category contains n or more objects. There's another form of the pigeonhole principle; we'll come to it eventually.

We're now ready to look at Ramsey's Theorem, version 2: For all m and $n \geq r \geq 1$, $\Re(m,n,r)$ exists and satisfies the relation

$$\Re(m,n,r) \leq \Re(\Re(m,n-1,r),\Re(m-1,n,r),r-1) + 1$$

for all m and $n > r > 1$.

Before proving this theorem, let's check it for the case $r = 2$. According to the theorem, $\Re(m,n,2) \leq \Re(\Re(m,n-1,2),\Re(m-1,n,2),1) + 1$. According to our formula for $\Re(m,n,1)$, the right-hand side evaluates to $[\Re(m,n-1,2)+\Re(m-1,n,2)-1] + 1$, so this checks against the first version of the theorem. How can we prove this new theorem in general?

Let $p = \Re(m-1,n,r)$ and $q = \Re(m,n-1,r)$, and choose any N such that $N \geq \Re(p,q,r-1)+1$. Suppose S is a set of N elements, and that the r-subsets of S have been colored solid and dotted. Pick any element v in S and consider the r-subsets that contain v. Each of them corresponds to an $(r-1)$-subset of the $N-1$ elements that do not include v. (Let S_0 represent the subset $S-\{v\}$, which is the same as S except that v has been removed.) A coloring of the r-subsets of S that include v corresponds to a coloring of the $(r-1)$-subsets of S_0 that do not include v. Since $N-1 \geq \Re(p,q,r-1)$, we know that either there is a subset of p elements of S_0, all of whose $(r-1)$-subsets are solid, or else there is a subset of q elements of S_0, all of whose $(r-1)$-subsets are dotted. The two cases are similar; consider the first. In this case we have a set containing $p = \Re(m-1,n,r)$ elements, each of whose $(r-1)$-subsets is solid. This corresponds to the coloring of the r-subsets of S in which each $(r-1)$-subset of the p elements, when taken together with v, forms an r-subset that is solid. Meanwhile, by the definition of $\Re(m-1,n,r)$, there must be either an $(m-1)$-subset of the p, all of whose r-subsets are solid, or else an n-subset all of whose r-subsets are dotted. In the latter case, we're done. In the former case, we take the $m-1$ elements together with v, and thereby find an m-subset all of whose r-subsets are solid.

Finally, let's generalize to include the situation where there are more than two colors available. We define $\Re(n_1,n_2,\ldots,n_t,r)$ to be the minimum integer such that, if each r-subset of a set containing $N \geq \Re(n_1,n_2,\ldots,n_t,r)$ elements is colored with one of t colors, then there is some i such that some n_i-subset has all its r-subsets colored using the ith color.

One more time! This time it's Ramsey's Theorem, version 3: For all $n_1, n_2, \ldots, n_t \geq r \geq 1$, $\Re(n_1,n_2,\ldots,n_t,r)$ exists and satisfies the relation

$$\Re(n_1,n_2,\ldots,n_t,r) \leq \Re(\Re(n_1,n_2,\ldots,n_{t-1},r),n_t,r)$$

for $t > 2$.

To prove this, we group the colors into two categories. The first category contains the first $t-1$ colors and the second category the last color. Given a set of $N \geq \Re(\Re(n_1,n_2,\ldots,n_{t-1},r),n_t,r)$ elements whose r-subsets have been colored using t colors, we take every r-subset that has been colored using any of the first $t-1$ colors and color it "solid" instead. By the definition of the two-color case, we know that either there is an n_t-subset all of whose r-subsets are dotted, or else there is a subset of $\Re(n_1,n_2,\ldots,n_{t-1},r)$ elements all of whose r-subsets have been colored using only the first $t-1$ colors. The theorem follows by induction.

For instance, we can get the following bound on the value of $\Re(3,3,3,2)$:

$$\Re(3,3,3,2) \leq \Re(\Re(3,3,2),3,2) = \Re(6,3,2) = 18.$$

(The earlier versions of the theorem told us only that $\Re(6,3,2)$ was no greater than 21, but it has in fact been found that it is 18, so we can get a slightly better bound on the value of $\Re(3,3,3,2)$.) The actual value of $\Re(3,3,3,2)$ is known to be 17. There'll be more to say about that on the final exam.

As a special case, we find $\Re(2,2,2,\ldots,2,1) = t+1$ (where there are t colors). This is another case of the pigeonhole principle: if we distribute $t+1$ objects into t categories, then some category must contain two or more objects. Ramsey Theory is in a sense just a generalisation of the pigeonhole principle.

For people interested in further reading on Ramsey Theory, Tarjan listed three references: [Hall], pp. 54–57, [Ryser], pp. 38–46, and [Harary], pp. 15–17.

Before moving on to chapter 10, let's look at some applications of Ramsey Theory. (This material was not covered in the lecture. This section is based on some lecture notes prepared by Tarjan that he never found time to present.)

A finite semigroup is a finite set on which a binary associative operation is defined. The operation is generally referred to as "multiplication", though in fact it may not be the ordinary arithmetic multiplication operation. For instance, given any finite set S, we can take the collection of all subsets of S, together with the operation of set union, and that will form a finite semigroup. (We could also use set intersection as the semigroup operation.) We will use "·" to denote the operation of the semigroup. An idempotent is an element e such that $e \cdot e = e$. It is claimed that any finite semigroup must have an idempotent.

To prove this using Ramsey Theory, we let a be an arbitrary element of the semigroup. Let n be the size (also called the order) of the semigroup, i.e., the number of elements that are in the set, and let $N = \Re(3,3,3,\ldots,3,2)$, where there are n 3's. Consider forming the product of N copies of a,

$$\underbrace{a \cdot a \cdot a \cdot a \cdot a \cdot \ldots \cdot a}_{N},$$

which we shall denote by a^N. (Similarly, a^t is the product of t copies of a, for any t.) We take the complete graph K_N and color the edges using n colors, where each color corresponds to a distinct element of the semigroup. We color the edge linking vertices i and j (where $i<j$) using the color corresponding to the element a^{j-i}. According to Ramsey's Theorem, there must be a monochromatic triangle, i.e., there must be some i, j, and k ($i<j<k$) such that $a^{j-i} = a^{k-i} = a^{k-j}$. Define $e = a^{j-i}$. Since $a^{k-j} \cdot a^{j-i} = a^{k-i}$, we have $e \cdot e = e$. Thus e is the desired idempotent.

For an alternate proof, we consider the powers of a. Since the semigroup is finite, we must eventually find two powers that are the same element; that is, there must be some positive integers i and j such that $a^{i+j} = a^j$. This implies by induction that $a^{ki+j} = a^j$ for all positive k. Choose some k such that $ki > j$. Then $ki-j$ is positive, and thus we can multiply both sides of the equation by a^{kl-j}. We find that $a^{kl+j+kl-j} = a^{j+kl-j}$, which means $a^{2kl} = a^{kl}$ and $e = a^{kl}$ is an idempotent.

For another situation in which Ramsey Theory can, rather unexpectedly, be applied, we turn to plane geometry. A region in a plane is said to be <u>convex</u> if every straight line connecting two points in the region lies entirely within the region. It is claimed that, for any n, there is a number $N(n)$ such that any planar set of $N(n)$ points, no three in a straight line, contains a convex n-sided polygon (n-gon).

To prove this, we first prove two lemmas. Lemma 1 states that, given five points in the plane that have no three in a line, some four of the five points must form a convex quadrilateral. To see this, consider the <u>convex hull</u> of the five points. The convex hull of a set of points is defined to be the smallest convex polygon that includes or contains all the points. (Another way of looking at it is to draw all edges connecting pairs of points in the set; the convex hull is then the "outer face" of the resulting graph.) If the convex hull contains 4 or 5 points, as shown in the first two diagrams below, the lemma is clearly true. If it contains only 3 points, as shown on the right below, the other two points are on the inside of the triangle. These two points determine a straight line, and since none of the other points can be along this line there must be two of the triangle's points on one side of the line. These two points, together with the two interior points, determine the desired quadrilateral.

The second lemma states that, given n points with no three in a line, if all quadrilaterals determined by subsets of 4 points are convex, then the n points determine a convex n-gon. Again we consider the convex hull of the n points. It is by definition a polygon; say it has q sides. We can break it into $q-2$ triangles, as shown on the following page. None of the n points can lie inside any of the triangles, or there would be a concave quadrilateral determined by that point and the enclosing triangle. Nor can there be any points outside the q-gon, by the definition of the convex hull. Hence all n points are included on the q-gon; *i.e.*, $q = n$, and the n points thus determine a convex n-gon as stated.

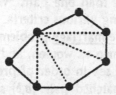

To prove the theorem, we pick $N(n) = \Re(5,n,4)$, where "solid" corresponds to "concave" and "dotted" to "convex". By the first lemma, no five points can have every 4-subset concave, so there must be n points with every 4-subset convex, which by the second lemma implies that the n points form a convex n-gon.

This proof, combined with Ramsey's Theorem, gives us an upper bound on $N(n)$. For some values of n, the minimum values of $N(n)$ are known. Specifically, $N(3) = 3 = 2+1$, $N(4) = 5 = 2^2+1$ (by the first lemma), and $N(5) = 9 = 2^3+1$. (It is an interesting exercise to try to find 8 points that do not include a convex pentagon.) It is unknown whether $N(n) = 2^{n-2}+1$ in general. Also unanswered is the following question: given n, is there a large enough $M(n)$ so that any planar set of $M(n)$ points, no three in a line, must contain n points defining a convex n-gon <u>with no other point inside</u>? It is known that $M(4) = 5$; the existence of $M(n)$ has not been proven for $n \geq 5$.

10   Matchings (Stable Marriages)

February 28. Given a set of men and a set of women, a __matching__ is a set of pairs, each pair containing one man and one woman, such that no person is in more than one pair. We shall be interested in finding matchings satisfying various criteria. The first problem we'll consider is called the __stable marriage__ problem. We assume that there are the same number of men as women, and that each person ranks the people of the opposite sex in order of preference. A matching is __stable__ if there is no unmatched pair {*a,b*} such that __both__ *a* and *b* prefer each other to their present partners. (If such a pair existed, they would run off together.) Though we speak of men and women, this is actually a rather facetious viewpoint; this problem is typically applied to relationships somewhat more pragmatic than marriage, such as roommate assignments, dormitory room assignments, and university admissions. Nevertheless, here we'll discuss the problem in terms of marriages between men and women.

For example, let's designate the men with lower-case letters and the women with upper-case. Suppose the preference lists are as follows:

$$a:\ A\,B\,C \qquad\qquad A:\ b\,a\,c$$
$$b:\ B\,A\,C \qquad\qquad B:\ c\,b\,a$$
$$c:\ A\,C\,B \qquad\qquad C:\ a\,c\,b$$

Suppose we take the matching {*aB, bC, cA*}. Is this stable? No, because *a* prefers *A* over *B*, and *A* prefers *a* over *c*, so *a* and *A* will run off together. Is there a stable matching for these people? Yes; here's one: {*aA, bB, cC*}. In this matching, *a* and *b* have their top choices and therefore will not want to run off with anyone else; *c* would prefer *A*, but *A* doesn't prefer *c*. This matching is therefore stable. So is this one: {*aC, bA, cB*}. It turns out that, no matter what the people's preferences are, there always exists at least one stable matching. We wish not only to prove this, but also to give a method for finding one. (This is thus a case of constructive combinatorics.)

The method we're about to present works something like this. The first man will propose to the first woman on his list. She, having no better offer at this point, will accept. The second man

G. Pólya et al., *Notes on Introductory Combinatorics*, Modern Birkhäuser Classics,
DOI 10.1007/978-0-8176-4953-1_10, © Birkhäuser Boston, a part of Springer
Science+Business Media, LLC 2010

will then propose to his first choice, and so on. Eventually it may happen that a man proposes to a woman who already has a partner. She will compare the new offer to her current partner and will accept whoever is higher on her list. The man she rejects will then go back to his list and propose to his second choice, third choice, and so forth until he comes to a woman who accepts his offer. (If this woman already had a partner, her old partner gets rejected and he in turn starts proposing to women further down his list.) Eventually everything gets sorted out. Now let's make this a bit more formal.

Algorithm:

Each person starts with no people "cancelled" from his or her list. People will be cancelled from lists as the algorithm progresses.

For each man m, do propose(m), as defined below.

propose(m):
Let W be the *first* uncancelled woman on m's preference list.
Do: refuse(W,m), as defined below.

refuse(W,m):
Let m' be W's current partner (if any).
If W prefers m' to m, then she rejects m, in which case:
    (1) cancel m off W's list and W off m's list;
    (2) do: propose(m). (Now m must propose to someone else.)
Otherwise, W accepts m as her new partner, in which case:
    (1) cancel m' off W's list and W off m''s list;
    (2) do: propose(m'). (Now m' must propose to someone else.)

Let's step through this algorithm using our earlier example. (This is the sort of thing best done at a blackboard, but we'll see what we can do.) We first do propose(a). a's top choice is A, so we do refuse(A,a). She has no current partner, so she accepts a. We proceed to the next man, b, whose first choice is B. She too accepts. Finally, we do propose(c), which in turn causes us to do refuse(A,c). Since A's current partner is a, and she prefers a over c, she rejects c. We cross A off of c's list, and cross c off of A's list, and do propose(c) again. This time, since A has been cancelled from his list, he proposes to C, who accepts. We have found a matching, and we have seen already that this particular matching is stable.

Let's look at a slightly more complex example. Consider the following preference lists.

$$a: D\,A\,C\,B \qquad\qquad A: a\,c\,d\,b$$
$$b: A\,C\,D\,B \qquad\qquad B: d\,a\,b\,c$$
$$c: C\,D\,B\,A \qquad\qquad C: c\,b\,d\,a$$
$$d: D\,A\,C\,B \qquad\qquad D: b\,d\,c\,a$$

Let's step through the algorithm using these lists. First a proposes to D, who accepts. Then b proposes to A, who accepts, and c proposes to C, who accepts. Finally, d proposes to D, and the fun begins: D accepts d's proposal and rejects a; a then proposes to A, who accepts a and rejects b; b proposes to C, but C is happy and rejects him, so b next proposes to D. Now D rejects d in favor of b, so d proposes to A. She rejects him, and he proposes to C, who also rejects him. Finally, d proposes to B, who accepts. Thus the final matching is $\{aA, bD, cC, dB\}$. Note that someone can end up with a despised partner (B was d's last choice) and the matching can still be stable. Note also that the algorithm can be run "the other way", with the women proposing to the men. Were we to do that in this particular example, we'd get the same solution with much less work.

Now let's prove a few things about this algorithm. First of all, *is* it an algorithm? That is, does it necessarily terminate? Yes it does, because no man ever proposes twice to the same woman. Next, is a matching generated? Yes: once a woman has a partner, she always has one (she may "trade" for a better one, but she'll never be without one). If there is an unmatched man, there must also be an unmatched woman. But an unmatched man keeps proposing until he has proposed to every woman, so he must eventually propose to the unmatched woman, and she will accept. So everyone ends up in the matching.

Now we come to the interesting part — is the matching stable? Well, suppose it isn't. Then there's some pair, say aB, who are not matched to each other, such that each prefers the other over his or her current partner. Let's assume a is paired with A and B with b. So a prefers B to A and B prefers a to b. Since B appears ahead of A on a's list, and a ended up proposing to A, we know that a must have proposed to B at some point. So why did B reject a? The only reason for B to reject a would be if she were paired at that point with someone she preferred over a. But if this were the case, she couldn't possibly have ended up with b as her partner, since each woman's partner can only improve.

Next, we wish to demonstrate that the solution found by the algorithm is _male optimal_, _i.e._, that no man can do any better than he does in the matching found by the algorithm. Since each man ends up matched with the first woman on his list who hasn't been cancelled, this is the same as saying that every time a woman W gets cancelled from a man m's list, it implies that _no_ stable matching includes the pair mW. Consider one such cancellation. W has proposals from m and m', and she rejects m. Suppose the pair mW occurs in some stable matching. In this matching, m' is paired with some other woman, W'. If W precedes W' in m''s list, then m' and W prefer each other over their assigned partners, and the matching isn't stable. Meanwhile, if W' precedes W in m''s list, then W' was cancelled off m''s list by the algorithm, and so we have yet another cancellation that must be in our hypothetical matching. Since each cancelled pair included in the matching implies that another such pair must also be in the matching, we have by induction that there must be an arbitrarily large number of pairs, cancelled by the algorithm, that are in this new matching. But that's impossible, because there aren't an arbitrarily large number of men and women in the sets. Hence our initial assumption, that there was at least one cancelled pair mW that could occur in some stable matching, must be wrong. Hence the matching found by the algorithm is male optimal.

Finally, we wish to demonstrate that the solution found by the algorithm is _female pessimal_, _i.e._, that no woman does worse in any other stable matching. Suppose some woman W is paired, in some stable matching S, with a man m' whom she ranks lower than the partner m assigned to her by the algorithm. Then W prefers m over m'. But, by male optimality, the man m prefers W to whomever he gets in the matching S. Hence S is unstable.

Pólya raised the question of whether there were any sort of global criteria of "goodness" which could be used to decide which of two matchings was "better". Such a criterion would presumably rate any stable matching as being better than any unstable one, and would rate stable matchings that were female pessimal (or male pessimal) lower than stable matchings that were more "balanced".

Such a measurement would indeed be nice to have, but none such seems to exist in general. We'll get back to this issue later.

What about partial preference lists? That is, suppose that for each person there are some people they dislike so intensely that they'd rather remain single than be paired with those people. In this case there need not be a solution (obviously, since one person could refuse to marry *anybody*). Given a collection of partial preference lists, we wish to determine whether there is in fact a stable matching in which all the men and women are paired.

March 2. To help solve this problem, we introduce one additional "dummy" man and one "dummy" woman. We'll denote them by m and W. For each man, we add W at the end of his partial list, followed by all the "real" women not already on his list. Similarly, we add m to each woman's list, followed by all real men not on the list. We also create preference lists for the two dummies. The order of the real men and women on these two lists is irrelevant; the only important feature is that W is m's <u>last</u> choice, and m is W's last choice. For instance, given the partial lists

$$a: D\,A \qquad\qquad A: a\,c$$
$$b: A\,C\,D \qquad\quad B: d\,a\,b$$
$$c: C \qquad\qquad\quad C: c$$
$$d: D\,A\,C\,B \qquad D: b\,d\,c\,a$$

we would transform them into the following:

$$a: D\,A\,W\,B\,C \qquad A: a\,c\,m\,b\,d$$
$$b: A\,C\,D\,W\,B \qquad B: d\,a\,b\,m\,c$$
$$c: C\,W\,A\,B\,D \qquad C: c\,m\,a\,b\,d$$
$$d: D\,A\,C\,B\,W \qquad D: b\,d\,c\,a\,m$$
$$m: A\,B\,C\,D\,W \qquad W: a\,b\,c\,d\,m$$

By our earlier analysis we know that there must be a stable matching for this set of lists, since now everybody lists a complete set of preferences. However, this stable matching might have someone matched to a person to whom he or she refuses to be married. We claim there is a complete, stable matching for the original problem if and only if there is a stable matching for the new problem in which m is paired with W.

Suppose we have a stable matching that includes the pair mW. Removing m and W can't make the matching unstable, since doing so can only reduce the number of possible alternatives for everyone. So all we need show is that no person prefers being single over being paired with his or her partner. Suppose some man a prefers being single to being married to his partner, A. Then a prefers W over A. But W prefers a to m, so this wasn't really a stable matching to begin with. The same argument applies if there is some woman who would prefer being single. The "only if" part of the claim is trivial and is left as an exercise.

We can go one step further and claim that, if some stable matching for the new problem includes the pair mW, then they all do. This follows directly from the male optimal solution, which we can find using our algorithm. If this solution includes the pair mW, then all stable matchings do, since m can't do any better, and W was his last choice. On the other hand, if the male optimal solution doesn't include mW, then no stable matching does, since this solution is female pessimal, implying that W can't do worse than whomever she's got in this matching.

Finally, suppose each person ranks all the others? (This is the sort of thing that might arise in, say, roommate assignments, since any person might be paired with any other.) For instance, consider the following preferences:

$$A: B\ C\ D$$
$$B: A\ C\ D$$
$$C: A\ D\ B$$
$$D: C\ A\ B$$

which have the complete, stable matching: $\{AB, CD\}$. Unfortunately, most of the results we proved regarding the bipartite case ("bipartite" means that there were two independent sets, and each pair included one element from each set) no longer apply. In particular, there need not be a stable matching, and there is no efficient method known for finding one if it exists.

Tarjan assigned as homework the problem of finding a set of preference lists for four people such that there is no stable matching. This isn't too difficult to do. Some people even managed to prove the uniqueness of the answer. Such a set of lists is the following:

$$A: B\,C\,D$$
$$B: C\,A\,D$$
$$C: A\,B\,D$$
$$D: \text{(arbitrary)}$$

Any set of lists of this form (that is, equivalent except for some interchanging of names) will have no stable matching, and these are the only such lists for four people. (We won't bother to prove that here.) There are 48 "unstable" sets of preference lists, out of a total of $(3!)^4 = 1296$ possible sets.

11 | Matchings (Maximum Matchings)

March 2. In the preceding chapter we mentioned the problem of devising a "global criterion" for deciding whether one matching is better than another. We noted that no such criterion appears to exist in general. Nevertheless, there are some cases where simple criteria suffice. For instance, if each possible pairing is either permitted or not permitted, with no other relative preferences given, then we have no trouble. This is known as the maximum matching problem.

Stated precisely, the maximum matching problem is this. (As in chapter 10, we'll discuss it in terms of men and women. A typical "real-life" application might involve assigning people to jobs. See also the "busing problem" mentioned on page 87.) We are given a set M of men and a set W of women. We are also given a set of "legal" pairs (a,b), where a is in M and b is in W. In this description, "legal" may be read "compatible"; *i.e.*, the pairs indicate men and women who can get along together. In the people/jobs situation, the legal pairs would be people a who are able to do job b. We restrict our definition of matching from chapter 10 so that we consider only legal pairs: A matching is a subset of the legal pairs such that each person is in at most one pair. A maximum matching is a matching containing as many pairs as possible. We are interested in finding (*i*) the size of (number of pairs in) this maximum matching, (*ii*) the matching itself, and (*iii*) as a special case, a means of determining easily whether all the elements of the smaller set can be matched.

For example, suppose there are four men and five women. We'll denote the men by the symbols x_1 through x_4 and the women by y_1 through y_5, and we'll show the legal pairs by drawing a graph in which an edge between x_i and y_j means (x_i,y_j) is a legal pair. Suppose the legal pairs are as shown on the left at the top of page 136. We can obtain the matching indicated by the jagged edges in the diagram on the right. Each jagged edge corresponds to selecting the pair consisting of the endpoints of that edge. Thus the matching is $\{(x_1,y_2), (x_2,y_1), (x_3,y_4), (x_4,y_5)\}$. This is obviously the maximum possible, since all the men are matched. This maximum is not unique; for instance, we could instead have matched x_4 with y_4.

G. Pólya et al., *Notes on Introductory Combinatorics*, Modern Birkhäuser Classics, DOI 10.1007/978-0-8176-4953-1_11, © Birkhäuser Boston, a part of Springer Science+Business Media, LLC 2010

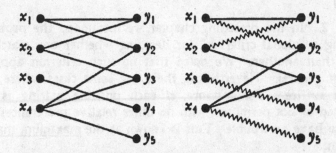

   This sort of graph is called a <u>bipartite</u> graph; the vertices of
such a graph can be partitioned into two sets such that no edge joins
two vertices from the same set. We can see from the above example
that the problem of finding a maximum matching corresponds to the
problem of finding a maximum set of edges in a bipartite graph
such that no two edges share a vertex. We can also transform the
maximum matching problem into one involving matrices. Given a
matrix in which each element is either 0 or 1, we wish to find a
maximum number of 1's with no two of them in a single row or
column. To transform a maximum matching problem into this form,
we create the matrix in which each row is associated with some x_i
and each column with some y_j, and let a 1 in the matrix indicate that
(x_i, y_j) is a legal pair. Our example yields the matrix shown below, in
which the parentheses indicate the set of ones corresponding to the
maximum matching designated by the jagged edges in the earlier
diagram.

	y_1	y_2	y_3	y_4	y_5
x_1	1	(1)	0	0	0
x_2	(1)	0	1	0	0
x_3	0	0	0	(1)	0
x_4	0	1	1	0	(1)

   Yet another problem equivalent to the maximum matching
problem is the following: given a collection of subsets of some set,
find a maximum set of distinct elements such that each element is
contained in a different subset. For instance, let's take the problem
corresponding to our favorite example. We let our collection of

subsets include one subset for each x_i, and the subset corresponding to any particular x_i will consist of those y_j that form legal pairs with x_i. Thus we have four subsets: $\{y_1, y_2\}$, $\{y_1, y_3\}$, $\{y_4\}$, and $\{y_2, y_3, y_5\}$. The maximum matching we've been using as a solution in the previous examples corresponds to selecting y_2 from the first set, y_1 from the second, y_4 from the third, and y_5 from the fourth. When, as in this case, the selected elements include one from each set (as opposed to leaving some sets unrepresented), the set of elements selected is called a <u>system of distinct representatives</u>. This is often abbreviated to SDR.

Since all of these classes of problems—maximum matching, disjoint edges, matrix element selection, and SDR—are equivalent (we haven't actually shown that, but the transformations described can obviously be performed in either direction), a method for solving any one of them can be applied to get solutions to the others. So we'll look at the problem of finding disjoint edges in a bipartite graph. How do we go about finding the largest such set of edges? We start by defining a few more terms. We'll use "matching" to refer to the set of edges selected, since any such selection corresponds to a matching in the original formulation of the problem.

A <u>free vertex</u> is one that is not contained in any edge of the matching. A <u>path</u> in a graph (we don't seem to have defined this term before) is a sequence of edges $v_1 v_2$, $v_2 v_3$, $v_3 v_4$, ..., $v_{k-1} v_k$. (While we're at it, we'll define a <u>cycle</u> (also called a <u>circuit</u>; we defined it informally on page 80) to be a path in which the last vertex, v_k, is the same as the first, v_1. We'll be talking more about cycles later.) A path is allowed to go through a vertex or vertices more than once; if it doesn't it is called a <u>simple path</u>. (Some texts use the terms "path" and "walk" instead of "simple path" and "path". We will occasionally use "path" to refer to a simple path when the non-repetition of vertices is obvious or unimportant.) Getting back to the problem at hand, we define an <u>alternating path</u> as a simple path consisting of alternating matched and unmatched edges. If an alternating path connects two free vertices, it is called an <u>augmenting path</u>. If an augmenting path exists, the size of the matching can be increased by one by switching the matched and unmatched edges along the path, as shown on the following page. (As before, jagged edges indicate edges in the matching.)

augmenting path   increased matching

Note that we needn't have restricted ourselves to simple paths, but if any vertex were repeated in an augmenting path we could remove the cycle thus formed and still have an augmenting path, so it's easier to consider only simple paths from the start.

This leads us to the following method for finding a maximum matching. This isn't really an algorithm, due to the vagueness of the second step.

> *Step 1:* Begin with any matching (*e.g.*, the empty set).

> *Step 2:* Look for an augmenting path. If one is found, increase the size of the matching accordingly. Repeat this step until no augmenting path is found.

We'll ignore for the moment the problem of finding augmenting paths. First let's prove that this method does indeed yield the maximum possible matching. Suppose it doesn't; suppose the real maximum is larger. Let M be a matching generated by the above method, and let M_0 be any maximum matching. We'll denote the size of M (the number of edges in M) by $|M|$. We want to show that, if $|M| < |M_0|$, there must be an augmenting path in M.

We define $M \oplus M_0$ by

$$M \oplus M_0 = (M \cup M_0) - (M \cap M_0)$$
$$= (M - M_0) \cup (M_0 - M),$$

where the notation $X-Y$ signifies all elements contained in X but not in Y. Thus $M \oplus M_0$ is all edges that are in either M or M_0 but not both. Since, in a matching, each vertex is incident to at most one edge, we know that each vertex is incident to at most *two* edges of $M \oplus M_0$. Hence the edges in $M \oplus M_0$ form only simple paths and cycles. Furthermore, if some vertex is incident to two edges, one of the edges must be in M (and not in M_0) and the other must be in M_0 and not in M. Thus the paths and cycles in $M \oplus M_0$ are *alternating* paths and *alternating* cycles in M. (Edges are alternately in M and not in M.) For any path or cycle, there are thus four possible forms, as shown below. Each of these could be extended to any length, but the basic forms would be the same: (1) a path beginning and ending with edges in M, (2) a path beginning and ending with edges in M_0, (3) a path beginning with an edge in M and ending with an edge in M_0 (the inverse case is achieved by following the path in the opposite direction), or (4) a cycle.

(1) M M_0 M M_0 M

(2) M_0 M M_0 M M_0

(3) M M_0 M M_0 M M_0

(4) M M_0 M_0 M M_0

We can partition the edges of $M \oplus M_0$ into simple paths and cycles of the above forms. We make each path as long as possible; that is, we don't break any path or cycle into two smaller paths. Note that paths of type (1) have more edges in M than in M_0, and that paths of type (3) and cycles of type (4) have exactly as many edges in M as in M_0. Only paths of type (2) have more edges in M_0 than in M. But we're assuming M_0 contains more edges than does M. Since $M \cup M_0$ contains all edges that are in either matching, it too contains more edges in M_0 than it does edges in M. (It may include, of course, some edges that are in both sets, and that therefore count toward both tallies.) Meanwhile, $M \cap M_0$ contains only those edges that are in both sets, and therefore obviously includes as many edges from M as from M_0. Thus, when we remove these edges from $M \cup M_0$, the resulting set (which is $M \oplus M_0$) must still contain more edges from M_0 than from M. Hence $M \oplus M_0$ must contain at least one path of type (2). This path is an augmenting path in M (the vertices at each end must be free or the path would have been made longer).

As an illustration, let's suppose M is the three-edge matching shown on the left on the following page, and M_0 is the maximum matching shown in the center (in which the edges included in the matching are drawn as dotted, rather than jagged, so that the two types are distinguishable in the third diagram). Then $M \oplus M_0$ is the graph shown on the right, and $x_1 y_2 x_4 y_5$ is seen to be a path of type (2) in $M \oplus M_0$, hence an augmenting path in M.

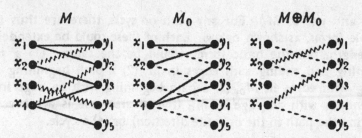

March 7. Now that we've shown that the augmenting path method will indeed find maximum matchings, let's return to the question of how to find the augmenting paths. There is an algorithm for doing this; it is called the <u>labelling algorithm</u> because it labels the men and women as they are reached via alternating paths. The labels show not only that the people have been reached, but also where the path came from that reached them. The rules for labelling vertices of the graph are shown below. Labels are written inside brackets.

(*i*)   Label any unlabelled free man with the label [–], indicating the beginning of an alternating path.

(*ii*) If b is an unlabelled woman joined by an unmatched edge to a labelled man a, then label b with [a].

(*iii*) If a is an unlabelled man joined by a matched edge to a labelled woman b, then label a with [b].

Notice that these rules take advantage of the bipartite nature of the graph. We'll see later how the problem becomes more difficult when the graph is not known to be bipartite. In the bipartite case, the labelling algorithm for finding augmenting paths is:

Apply the above labelling rules repeatedly until either
     (1) a free woman is labelled, or
     (2) nothing more can be labelled.
In case (1), an augmenting path exists, and the number of edges in the matching can be increased accordingly. (The path can be determined by tracing back through the labels.)
In case (2), there is no augmenting path and the matching is therefore maximum.

To see better how this algorithm works, let's apply it to the non-maximum matching M shown on the left at the top of page 140. There is only one free man, namely x_1, which by rule (i) is labelled [–]. Rule (ii) then permits us to label y_1 and y_2 with the label $[x_1]$ (see leftmost diagram below). By rule (iii) we can then label x_2 with $[y_1]$ and x_4 with $[y_2]$. At this point (rightmost diagram below) we could label y_3 with either $[x_2]$ or $[x_4]$ using rule (ii), or we could label y_5 with $[x_4]$. We choose to label y_3 with $[x_4]$; having labelled a free woman, we are done (bottom diagram). We can find the augmenting path by starting at the labelled free woman, y_3, and looking at her label, which is $[x_4]$. We then look at x_4's label, which is $[y_2]$. The label on y_2 is $[x_1]$, and x_1's label is [–], indicating the beginning of the path. The augmenting path is therefore $x_1 y_2 x_4 y_3$.

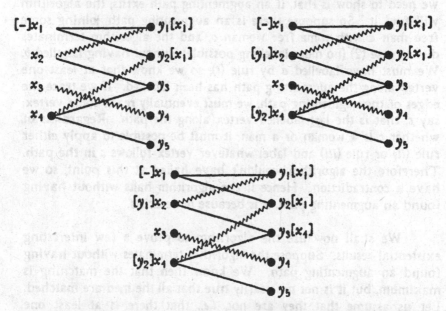

If we increase the matching using this augmenting path, we obtain the matching shown on the left on the following page. This matching is maximum, and if we try applying the algorithm it halts immediately, because there are no free men to be labelled using rule (i). We could also run the algorithm "backwards", treating the y's as men and the x's as women. In this case, we would label y_5 with [–], and the algorithm would eventually finish with the graph labelled as shown on the right. Since nothing more can be labelled, and no free x has been labelled, the algorithm tells us the matching is maximum.

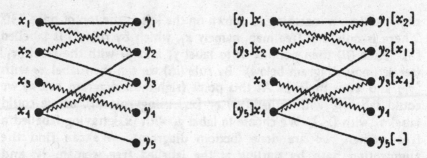

It should be obvious from the labelling rules that, if the algorithm terminates by labelling a free woman, it has indeed found an augmenting path. To prove that the algorithm works, then, all we need to show is that, if an augmenting path exists, the algorithm will find it. So suppose there is an augmenting path joining some free man a with some free woman b, and the algorithm terminates due to case (2) (no more labelling possible) without having labelled b. We must have labelled a by rule (*i*), so we know that at least one vertex along the augmenting path has been labelled. If we trace the edges of the augmenting path, we must eventually reach some vertex, say c, that is the <u>last</u> labelled vertex along the path. Regardless of whether c is a woman or a man, it must be possible to apply either rule (*ii*) or rule (*iii*) and label whatever vertex follows c in the path. Therefore the algorithm wouldn't have halted at this point, so we have a contradiction. Hence if the algorithm halts without having found an augmenting path, it is because none exists.

We shall now use the algorithm to prove a few interesting existential results. Suppose the algorithm terminates without having found an augmenting path. We know then that the matching is maximum, but it is not necessarily true that all the men are matched. Let us assume that they are not, *i.e.*, that there is at least one unmatched man. Let F be the set of free (unmatched) men, X the set of matched but labelled men, and Y the set of labelled women. We observe that, since no augmenting path was found, only matched women can be labelled. Furthermore, the only way a matched man can be labelled by the algorithm is for the woman to whom he is matched to be labelled also; conversely, if a woman is labelled then the man to whom she is matched must be labelled also. Hence we conclude that $|X| = |Y|$. We also observe that $|F| \geq 1$.

We define, for any set Z of men, the function $B(Z)$ to be the set of women who are joined by edges to one or more men in Z. For instance, in our favorite example, if Z is the set $\{x_1, x_2\}$, then $B(Z)$ is the set $\{y_1, y_2, y_3\}$. We claim that $B(F \cup X) = Y$. The justification for this is simple. Consider any edge joining a man m in $F \cup X$ to some woman w; we wish to show that w must be in Y, *i.e.*, that she must be labelled. There are two cases to consider. If the edge joining m and w is in the matching, then m cannot be in F and must therefore be in X. Thus m is labelled but not free, and the only way such a man can be labelled is by applying labelling rule (*iii*). Hence m must be joined by a matched edge to a labelled woman. But there can be at most one matched edge for any given man, so the edge leading to w is the only matched edge for m, and w must be labelled. On the other hand, if the edge joining m and w is <u>not</u> in the matching, then since m is labelled we know by rule (*ii*) that w must be labelled.

One more observation, after which we'll be ready to state a theorem. Since $B(F \cup X) = Y$, and F and X are disjoint, and $|F| \geq 1$, we know that $|B(F \cup X)| = |Y| = |X| < |F \cup X|$. Now for the theorem. It is claimed that the men can be matched *completely* if and only if, for *all* sets Z of men, $|Z| \leq |B(Z)|$. The proof is simple. If there exists a set Z such that $|Z| > |B(Z)|$, then it is clearly impossible to match all of the men in Z, since each must be matched to a different woman in $B(Z)$. Conversely, suppose it is not possible to match all of the men. The labelling algorithm then produces sets F, X, and Y as defined earlier, and if we take Z to be $F \cup X$ we know that $|Z| > |B(Z)|$. The theorem is thus proved.

We next define the <u>deficiency</u> of a set Z of men to be $|Z| - |B(Z)|$ if this is greater than zero, and zero otherwise. The <u>maximum deficiency</u> is the maximum, taken over all sets Z of men, of the deficiency of Z. We claim (here comes another theorem) the size of a maximum matching equals the number of men minus the maximum deficiency.

This theorem is obviously true if the maximum deficiency is zero, since in that case we know from the previous theorem that we can get a complete matching. Also, since a set of men Z with a positive deficiency $|Z| - |B(Z)|$ cannot have more than $|B(Z)|$ of its men matched in any matching, it is clear that the size of a maximum matching must be less than or equal to the total number of men minus the maximum deficiency. We need only show that the size of the matching is also *greater than* or equal to this number; this will imply equality. From our earlier analysis we know that, in any maximum matching, $|FUX| - |B(FUX)| = (|F|+|X|) - |Y| = |F|$. Since the maximum deficiency is, by definition, at least as great as the deficiency of any particular set, we know that it is greater than or equal to $|FUX|-|B(FUX)|$ (since this is simply the deficiency of the set FUX). Hence the number of unmatched men (*i.e.*, $|F|$) is less than or equal to the maximum deficiency. The theorem follows.

A few pages back we defined what we meant by a system of distinct representatives (SDR). Since finding an SDR is equivalent to finding a complete matching, it makes sense that there should be for SDRs a parallel to the theorem regarding complete matchings. This parallel theorem is known as <u>Hall's Theorem</u>, and it states that an SDR exists for a collection of subsets if and only if, for <u>every</u> k, any collection of k of the subsets contains at least k distinct elements. For example, suppose we look at the SDR problem corresponding to the matching problem we've been using. There are four subsets, each of which consists of the women who are joined to a given man. The sets are therefore $\{y_1,y_2\}$, $\{y_1,y_3\}$, $\{y_4\}$, and $\{y_2,y_3,y_5\}$. It is easy to see that each set contains at least one element, each pair of sets contains at least two different elements, and so on. Since the ith set is equivalent to $B(\{x_i\})$, it is clear that Hall's Theorem is equivalent to the theorem we've already proven, which stated that a complete matching exists if and only if $|Z| \leq |B(Z)|$ for all sets of men Z.

Continuing the process of referencing the second previous theorem, Tarjan next presented a corollary to the theorem that related the size of a maximum matching to the maximum deficiency. The corollary states that, if every man is joined to <u>at least</u> k women, and every woman to <u>at most</u> k men, then the men can be completely matched. To see this, we let Z be any set of men. Since each man is joined to at least k women, there must be at least $k|Z|$ edges coming from men in Z. To how many different women do these edges lead? Each woman can account for at most k edges, so there must be at least $(k|Z|)+k$ women involved. Hence $|B(Z)| \geq k|Z|+k = |Z|$.

From the preceding corollary we can derive another, which we shall state without proof. (The proof is trivial and is left as an exercise.) If there are fewer men than women, and each man is joined to the same number of women, and each woman is joined to the same number of men, then the men can be completely matched.

We can use these results to prove Sperner's Theorem, which we encountered in chapter 7 (though we didn't give it a name then). It concerned the problem of finding the largest collection of subsets of a given set, subject to the condition that none of the subsets should be a subset of another of the subsets. (In chapter 7 we used Pólya's terminology and called such sets "disconnected". Here, to avoid possible conflict with the term as it applies to graphs, we'll switch to Tarjan's terminology and call the sets "incomparable".) Sperner's Theorem states that the maximum is achieved by taking half the size of the original set and letting the collection consist of all subsets of that size. Thus there are $\binom{n}{\lfloor n/2 \rfloor}$ subsets in the collection. For instance, if the original set is $\{a,b,c,d\}$, it is not possible to have more than $\binom{4}{2} = 6$ subsets without one of them being contained in another, and it *is* possible to have exactly 6 such subsets, namely $\{a,b\}$, $\{a,c\}$, $\{a,d\}$, $\{b,c\}$, $\{b,d\}$, and $\{c,d\}$.

You might not expect maximum matchings would have any application to this problem, but it turns out we can use some of the results we've just proven to prove Sperner's Theorem. Though this proof is non-trivial, it is considerably simpler than most (perhaps even all) other proofs of the theorem. We shall consider any arbitrary collection of incomparable subsets and show that it can be systematically modified, without decreasing its size (i.e., the number of subsets), until it contains only subsets of size $\lfloor n/2 \rfloor$.

Let i be the size of the smallest subset, and let j be that of the largest. There are three cases (of which the first two are not mutually exclusive): (1) $i < \lfloor n/2 \rfloor$, (2) $j > \lfloor n/2 \rfloor$, or (3) $i = j = \lfloor n/2 \rfloor$. Suppose $i < \lfloor n/2 \rfloor$. We construct a matching problem in which the "men" are all possible subsets of size i, the "women" are all possible subsets of size $i+1$, and a set x of size i is joined by an edge to a set y of size $i+1$ if and only if x is a subset of y. We will show that there is a complete matching for the sets of size i; we can then replace each incomparable set of size i by its matching set of size $i+1$. In so doing, we cannot affect the incomparability of the sets, since we are only increasing the size of the smallest sets. (Adding an element to a set cannot make it a subset of another set, and we will leave no smaller sets which could be subsets of this one.) How do we know a complete matching exists? Each set x of size i is joined to $n-i$ sets of size $i+1$ (one for each element not in x), while each set y of size $i+1$ is joined to $i+1$ sets of size i (one for each element in y).   Since $i < \lfloor n/2 \rfloor$, which implies $i \leq (n-1)/2$, we know $n-i \geq (n+1)/2 \geq i+1$, so the first corollary from page 144 tells us that a complete matching must exist. (The second corollary could also be applied.) The case $j > \lfloor n/2 \rfloor$ is handled similarly and is left as an exercise.

As homework, Tarjan assigned a problem involving what is called a <u>system of simultaneous representatives</u>. Given a set S that has been partitioned into n subsets in two different ways, thus:

$$S = A_1 \dot\cup A_2 \dot\cup A_3 \dot\cup \cdots \dot\cup A_n = B_1 \dot\cup B_2 \dot\cup B_3 \dot\cup \cdots \dot\cup B_n$$

where the symbol "$\dot\cup$" denotes the union of disjoint sets), a system of simultaneous representatives (SSR) is a set of n distinct elements x_1, $x_2, \ldots, x_n$ that contains one element from each of the A_i and also contains one from each B_i. That is, the n elements form an SDR for the A_i, and the same n elements form an SDR for the B_i. The assignment was to prove that the following condition is necessary and sufficient for the existence of an SSR: for every k from 1 to n, no union of k of the A_i is contained in the union of fewer than k of the B_i.   For example, suppose S is the set $\{a,b,c,d,e,f\}$, $n = 3$, the A-sets are $\{a,b\}$, $\{c,d\}$, and $\{e,f\}$, and the B-sets are $\{a,b,c\}$, $\{d,f\}$, and $\{e\}$.   There is an SSR, namely $\{a,d,e\}$. But if instead the B-sets were $\{a,b,c,d\}$, $\{f\}$, and $\{e\}$, then there would be a union of two A-sets, namely $A_1 \cup A_2$, that was contained in a single B-set, namely B_1, so there would be no SSR.

This should have been a trivial assignment, but a remarkable number of students managed to turn it into an extremely complicated task. The necessity of the condition is easy to prove using any of a number of approaches, but proving sufficiency by, say, induction can be quite tricky. Few of those who tried such an approach managed to come up with a valid proof. On the other hand, haven't these people ever heard of the idea of using one result to prove another? We've proven a number of theorems already; why should we have to start from scratch every time we want to prove another result? Clearly, we shouldn't. To prove that the given condition is necessary and sufficient, we construct a matching problem as follows. Let the "men" be the sets $A_1, A_2, \ldots, A_n$, and let the "women" be the sets B_1,

$B_2, \ldots, B_n$. Two sets A_i and B_j are a legal pair if and only if their intersection $A_i \cap B_j$ is not empty; *i.e.*, the two sets have at least one element in common. If a complete matching exists, we can use it to construct an SSR by taking, for each pair (A_i, B_j) in the matching, any element from $A_i \cap B_j$ and placing it in the SSR. Since each of the A-sets and each of the B-sets occurs exactly once in a complete matching, this process yields n distinct elements forming an SDR both for the A-sets and for the B-sets. Conversely, suppose an SSR exists; we can use it to find a complete matching. By definition, the SSR is a set of elements $x_1, x_2, \ldots, x_n$ such that each x_k is contained in a different A_i from the others, and is also contained in a different B_j from the others. Thus we take the pair (A_i, B_j) to be in the matching; we know that no A-set or B-set can occur twice, so the matching must be complete.

We have shown that an SSR exists if and only if there is a complete matching for the A-sets and B-sets. By one of our earlier theorems, such a matching exists if and only if, for all collections Z of the A-sets, $|Z| \le |B(Z)|$. (Note that $|Z|$ is the number of A-sets included in Z, not the number of elements contained in those A-sets.) But $B(Z)$ is merely the collection of B-sets having any elements in common with any of the A-sets in Z; hence $|B(Z)|$ is the minimum number of B-sets that contain all of the A-sets in Z. So the necessary and sufficient condition stated in our earlier theorem is equivalent to the condition we're supposed to prove regarding the SSR. Q.E.D.

March 9. If we look at the matrix equivalent of a maximum matching problem, we encounter another theorem. This one is due to König and Egerváry, and it states that, in a matrix of zeros and ones, the maximum number of ones, no two in a line (row or column), equals the minimum number of lines needed to cover all the ones. Thus, if we consider our favorite example in its matrix form (as shown at the top of page 148), we find that the maximum number of ones, no two in a line, is four, as indicated by the parentheses in the diagram. It is clearly possible to cover all eight ones with four lines; just take the four rows. It is also clear that at least four lines are required. It is obvious that the minimum number of lines must be at least as great as the maximum number of pairwise noncollinear 1's, since no line can cover more than one of those 1's. To prove equality, we shall first place ourselves back on

	y_1	y_2	y_3	y_4	y_5
x_1	1	(1)	0	0	0
x_2	(1)	0	1	0	0
x_3	0	0	0	(1)	0
x_4	0	1	1	0	(1)

familiar ground by stating the equivalent theorem in terms of the disjoint edges problem on bipartite graphs. For graphs, the theorem tells us that the size of a maximum matching equals the minimum number of vertices needed to cover all the edges. (An edge is "covered" if either of its endpoints is among the selected vertices; it needn't have both its endpoints included.)

Let A and B be the sets being matched. (Each element of A corresponds to a row of the matrix; each element of B corresponds to a column.) Let m be the size of a maximum matching. We wish to show that there must be a set C of m vertices covering all the edges between A and B. If $m = |A|$, the theorem is obvious (let $C = A$). Suppose $m < |A|$. Let Z be a set of men with maximum deficiency. We know from our earlier theorems that

$$m = |A| - (|Z| - |B(Z)|).$$

Let $C = (A-Z) \cup B(Z)$, i.e., the set of all men not in Z plus all women joined to men in Z. The size of this set is, by the above formula, m. Now consider any edge xy. If x is contained in $A-Z$, then this edge is covered by C. Otherwise, x must be in Z, and thus y is in $B(Z)$, so the edge is still covered by C. Thus C covers all the edges, and the theorem is proved.

Before moving on, let's look briefly at the problem of finding a maximum set of disjoint edges in a *non-bipartite* graph. Most of our results from the bipartite case do not apply; however, it *is* still true that a matching (set of edges) is maximum if and only if there is no augmenting path. The problem is, how do we find augmenting paths? If we try using the labelling algorithm, we will run into trouble when we encounter odd cycles, that is, cycles containing an odd number of edges. To see why this is so, consider the example below. The vertices have been assigned letters so we can refer to them in the labelling procedure. Suppose we start by labelling a with [-] by rule (i). We can then label b with [a], c with [b], d with [c], and e with [d]. At this point (second diagram below), if we label h and i next, we're all set, because we can then label j with [i] and have an augmenting path from a to j. On the other hand, we could just as well label f and g instead (third diagram below). Having done this, we might then label i with [g], and now we cannot label j with [i] because this would not be an alternating path.

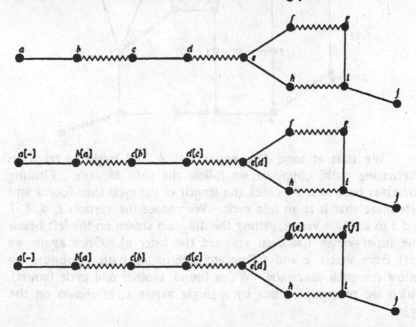

To handle the problems introduced by odd cycles, Tarjan presented something he called "Edmond's Incredible Shrinking Cycle Algorithm". (Edmond, as you might have guessed, is the inventor of the algorithm.) What this algorithm does is to apply the labelling scheme essentially as before, but whenever an odd cycle is detected it is replaced by a "super-vertex", i.e., a single vertex that represents all the vertices included in the cycle. Tarjan did not describe the algorithm in detail, and instead illustrated it by example. We'll do our best to reproduce that example in these notes. Consider the graph shown on the following page, in which the jagged edges as usual represent edges currently selected to be in the matching. This graph does contain an augmenting path, though it may not be obvious at first glance. It is the path *abcdklnmefgh*. Let's step through Edmond's algorithm and see how it manages to find this augmenting path.

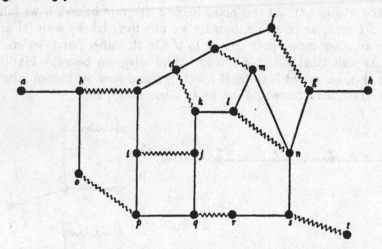

We start at some free vertex, say *a*, and begin to trace an alternating path. Suppose we follow the path *abcdkjic*. Finding ourselves back at *c*, we check the length of the cycle thus found and determine that it is an odd cycle. We reduce the vertices *c*, *d*, *k*, *j*, and *i* to a single vertex, getting the diagram shown on the left below (the super-vertex has been assigned the letter *u*). Once again we start from vertex *a* and follow any alternating path. Suppose we follow the path *abuemnlm*. We've found another odd cycle (*mnlm*), which we promptly replace by a single vertex *v*, as shown on the right.

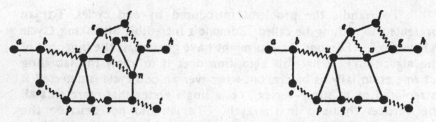

Back to vertex *a*, and this time we might happen to follow the path *abuevu*. We reduce the cycle *uevu* to a single vertex *w*, getting the diagram shown on the left on the following page, and proceed.

We follow the path *abwgfw*, and reduce the odd cycle *wgfw* to form a single vertex *x* (right). Now, starting once more from vertex *a*, we might find the even cycle *bxpob*, and the dead-end path *abxst*, but sooner or later we'll find the augmenting path *abxh*.

Having found this augmenting path in the reduced graph, we're still faced with the problem of deriving the augmenting path in the original graph. This is not as difficult as it might first appear. We trace along the augmenting path *abxh* until we come to a super-vertex. The only one on the path is *x*. We recall that *x* was formed by combining the three vertices *w*, *g*, and *f*. The augmenting path enters this cluster of vertices via *b* on one side and *h* on the other. Thus we know that the edges *bw* and *gh* must be part of the path. Since *w*, *g*, and *f* form an odd cycle, it must be possible to follow the cycle in one direction or the other to "link up" the path. It turns out that the direction to follow is via *f*. So we have converted the path *abxh* into the path *abwfgh*. We examine this new path for super-vertices, and find *w*. We expand it to reintroduce the vertices *u*, *e*, and *v*, and find that *b* enters this cluster via the edge *bu* while *f* enters via *ef*. Again, there must be one direction around the cycle forming an alternating path; it's the direction that goes through *v*. We now have the path *abuvefgh*. Expanding *u* and *v* in similar fashion produces the augmenting path we want.

Network flow problems are usually presented in terms of directed graphs. A directed graph is the same as the graphs we've been working with, except that each edge is assigned a direction. Another way of stating the distinction is that, whereas in an undirected graph each edge is a pair of vertices, in a directed graph each edge is an ordered pair of vertices.

In a network flow problem, each edge is assumed to have a capacity, indicating the maximum quantity (of whatever is flowing, such as oil or traffic) that can flow along that edge (in the direction of the edge). For instance, consider the graph below (left). The dotted lines will be explained later. The direction of the edges is indicated by arrows, and each edge is marked with its capacity. The vertex marked s is called the source, and the vertex marked t is the sink. The source is assumed to have an infinite supply of whatever it is that is flowing, and the sink has an infinite capacity. (It is easy to modify the graph (using additional vertices and edges) if we wish to impose limits on the source and sink, or have multiple sources and sinks, etc.) The objective of the network flow problem is to find the maximum flow from the source to the sink subject to the conditions that (*i*) flow through an edge must not exceed its capacity and (*ii*) flow at each vertex (except s and t) is conserved. The flow out of s is, of course, equal to the flow into t. Either of these flows is called the value of the flow. In the example below, the maximum flow is 3 "units", which can be achieved by letting the flow through the various edges be as shown on the right.

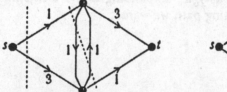 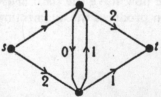

A cut is a line or set of lines that completely severs all connections between the source and the sink. Each of the two dotted lines shown in the diagram on the left above indicates a possible cut in this network. The capacity of a cut is the sum of the capacities of all edges crossing the cut from the source's side to the sink's. Thus

G. Pólya et al., *Notes on Introductory Combinatorics*, Modern Birkhäuser Classics,
DOI 10.1007/978-0-8176-4953-1_12, © Birkhäuser Boston, a part of Springer
Science+Business Media, LLC 2010

the capacities of the two cuts shown are 4 (for the cut that crosses only two edges) and 3 (for the other, which crosses three "forward" edges and one "backward" edge; the backward edge is ignored). There is a classic theorem in network flow theory, first formulated by Ford and Fulkerson. It is usually called the <u>Max Flow Min Cut</u> theorem, and it states (appropriately enough) that the value of the maximum flow is equal to the capacity of the minimum cut. As usual, we shall prove this theorem by constructing an algorithm, proving that it works, and then using it to prove the theorem.

Once again we shall employ an augmenting scheme. Before describing it, let's look at some examples to get an idea as to what the augmenting paths look like. Suppose, in the example we've just looked at, we started by finding an arbitrary simple path from s to t. In particular, suppose we take the path shown on the left below, and let one unit flow along this path from s to t. We could then apply either of the augmenting paths shown (center and right), among others. Note that the path on the right does not involve the edge running "down the page" from the upper vertex to the lower, but instead travels *backwards* along the upward edge. This is permitted because this edge currently has a positive flow assigned to it; by traversing it in the direction opposite that of the flow, we would *decrease* the flow assigned to that edge.

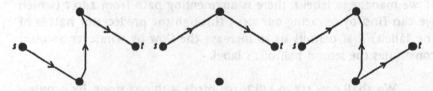

In general, an augmenting path from s to t consists of a simple path from s to t such that if an edge appears in the forward direction it is unsaturated (*i.e.*, the flow currently assigned to it is less than its capacity) and if it appears in the backward direction it has some positive flow. Armed with this definition of augmenting paths, we are ready to present an algorithm.

Step 1: Start with zero flow in all edges.

Step 2: Look for an augmenting path. If one is found, increase the flow accordingly. Repeat this step until no augmenting path is found.

As in the case of the algorithm for finding maximum matchings, we need to show that (i) if no augmenting path exists, the flow must be maximum, and (ii) there is an efficient way to find an augmenting path if one exists. We'll start with the latter. We'll use a labelling process; this time the labels will consist of two parts. The first part of the label of a vertex v will be the vertex that precedes v on the augmenting path that reaches v. The second part of the label will be the amount of additional flow that can be shipped from s to v along that path (regardless of whether this flow can reach all the way to t).

The labelling rules are as follows. (The notation $\underline{\min}(a,b)$ denotes the smaller of the two values a and b.)

(i)   Label the source $[-,\infty]$, indicating no predecessor and an infinite capacity.

(ii)   If some vertex v is labelled $[*,x]$ ($*$ is arbitrary), and $v \rightarrow w$ is an unsaturated edge with capacity c and current flow f, and w is not yet labelled, then label w with $[v,\underline{\min}(x,c-f)]$.

(iii) If v is labelled $[*,x]$ and $w \rightarrow v$ is an edge with positive flow f, and w is not yet labelled, then label w with $[v,\underline{\min}(x,f)]$.

If we manage to label t, there is augmenting path from s to t (which we can find by retracing our steps through the predecessor halves of the labels) that permits us to increase the flow by whatever amount constitutes the second half of t's label.

We shall now try to kill three birds with one stone, by proving that the labelling algorithm will always find an augmenting path if one exists, and that if none exists the flow is maximum, and that the value of the maximum flow equals the capacity of the minimum cut. It is clear from the definition of the capacity of a cut that the maximum flow cannot be greater than the minimum cut. The problem is proving equality. Suppose the labelling algorithm gets stuck; that is, it terminates without labelling t. Let X be the set of labelled vertices. Clearly s is in X and t is not. For each edge $v \rightarrow w$ such that v is in X and w is not in X, we know that the edge's capacity must equal its current flow. (The flow cannot exceed the capacity, and if it were less than the capacity we would have labelled w using rule (ii).) Similarly, if $w \rightarrow v$ is an edge with v in X and w not in X, then the flow through the edge must be zero (else w would have been labelled using rule (*iii*).)

We now examine the cut that separates X from the rest of the network. By our preceding observations, all edges crossing this cut toward t are saturated, and all edges crossing into X have no flow. Hence the current flow from X to the rest of the network, and thus the flow from s to t, is equal to the capacity of this cut. Since the capacity of this cut is at least as great as that of the minimum cut, and since the current flow cannot be greater than the maximum possible flow, it follows that the capacity of the minimum cut must be less than or equal to the maximum flow. This proves the Max Flow Min Cut theorem. Having found a cut with a capacity equal to the current flow, we know that the maximum flow cannot be any larger than the one we've got, so there can be no augmenting paths. Thus, if the labelling algorithm terminates without finding an augmenting path, it is because no such path exists. Furthermore, if no augmenting path exists, the algorithm must certainly terminate without finding one, and the cut described above proves that we have found a maximum flow. In other words, if no augmenting path exists the flow is maximum, and if an augmenting path *does* exist the labelling algorithm will find it.

Before moving on to chapter 13, we feel we should point out some of the limitations of the algorithm just presented. First, it doesn't necessarily work on networks in which the edges may have irrational capacities. In such networks it is possible the algorithm will find an infinity of augmenting paths that increase the flow by ever-decreasing amounts, such that the maximum flow is never achieved. In fact, one can construct networks in which the infinite sequence of ever-greater flows approaches a limit that is less than the true maximum! (For an example of such a network, see page 21 in [Ford-Fulkerson].) Even when the capacities are rational, it can happen that the algorithm takes an unduly long time to find the maximum flow. For example, consider the network shown on the left on page 156. The maximum flow is clearly 200000 units, and can be achieved by using a mere two augmenting paths—one going across the top and one along the bottom. On the other hand, there's no guarantee the algorithm will find those particular augmenting paths. It could instead find the path shown in the center diagram, which increases the flow by a single unit. The algorithm might then find the path shown on the right, traversing the central edge in a backward direction, thereby decreasing the flow in that edge to zero as the total flow is incremented by another unit. By always chancing to find one of these two paths (whichever applies), the algorithm would manage to increase the flow by only one unit per path. It would eventually find the maximum flow, but it would take it a while!

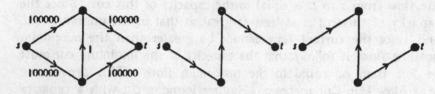

One can refine the augmenting algorithm so as to guarantee it always finds the maximum flow (even given irrational capacities) within a reasonable length of time. (Such refinements are among the topics discussed in one of Tarjan's more advanced courses.) For more information about network flow problems, see [Hall], [Harary], and [Ford-Fulkerson].

March 14. Hamiltonian and Eulerian paths and cycles come under the general heading of "de Bruijn sequences". The specific terms "Hamiltonian" and "Eulerian" are somewhat better known; hence this chapter has been named after them rather than de Bruijn.

Tarjan began the lecture by introducing two apparently quite unrelated problems. The first was something he called a <u>memory wheel</u>. This is a directed cycle in which each vertex is marked either '0' or '1', which contains all 2^k different sequences of length k. For example, shown below is a memory wheel for the case $k = 2$. If we start at the left and (as indicated by the direction of the edges) follow the circle clockwise, we get the sequence 00. If we start at the top, we

get 01. Starting at the right gives 11 and at the bottom gives 10. The problem is to determine, for some arbitrary k, the smallest memory wheel (*i.e.*, the shortest cycle) containing all binary sequences of length k. (The problem can also be extended to cases where more than two different digits are involved, but the particular case of binary sequences is of great interest in information theory.) Clearly, since there are 2^k different sequences, there must be at least 2^k different vertices in the cycle, since no vertex can be the starting point of more than one sequence. It turns out (as we shall prove later in this chapter) that it is always possible to find a memory wheel with exactly 2^k vertices. In fact (though we won't prove this), there are exactly $2^{(2^{k-1}-k)}$ different memory wheels with 2^k vertices.

The second problem was a classical problem known as "The Seven Bridges of Königsberg". The Prussian city of Königsberg (which has since been renamed Kaliningrad) has the Pregel river running through it, such that the city occupies both sides of the river as well as two islands. The islands are connected to the mainland (and to each other) by seven bridges, as shown in the somewhat

G. Pólya et al., *Notes on Introductory Combinatorics*, Modern Birkhäuser Classics,
DOI 10.1007/978-0-8176-4953-1_13, © Birkhäuser Boston, a part of Springer
Science+Business Media, LLC 2010

stylised diagram below. The question was, could a person leave home, take a walk, and return, crossing each bridge exactly once? It

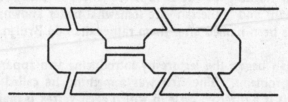

turns out this is impossible, but coming up with a characterisation of *why* it is impossible was a problem that stymied mathematicians for a long time until Euler managed to solve it.

We'll return to the Königsberg problem eventually, but first let's examine the memory wheel problem. It can be converted into a problem involving Hamiltonian cycles, which we shall now define.

A <u>Hamiltonian cycle</u> within a graph is a cycle that passes through each <u>vertex</u> of the graph <u>exactly once</u>. (The reason for the emphasis on "vertex" will become clear later.) The graph may or may not have directed edges. For example, consider the undirected graph shown on the left below, which is the graph formed by the vertices and edges of a regular dodecahedron (a regular solid with twelve pentagonal faces), distorted to be drawn on the plane of the paper. Though it may not be obvious at first glance, this graph has a Hamiltonian cycle, as shown in the diagram on the right.

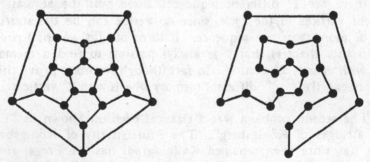

In general, it is obviously possible to determine whether an arbitrary graph contains a Hamiltonian cycle; simply try all possible sequences of vertices. If there are n vertices, this entails checking a mere $n!$ sequences. (Let us know when you're done.) No one knows an efficient way to determine whether an arbitrary graph has a

Hamiltonian cycle. (By "efficient" is meant any method taking time proportional to some polynomial in n, where n is the number of vertices in the graph.) On the other hand, no one has been able to prove that such an algorithm cannot exist. This is one of a large class of diverse problems that have been proven to be "equally hard", in the sense that if there exists a polynomial-time algorithm for any one of these problems, then there must be polynomial-time algorithms for them all. Anyone who finds an efficient algorithm, or proves that no such algorithm exists, for *any* of these problems, is guaranteed instant fame. But we digress.

In some cases, of course, it is possible to prove a graph has no Hamiltonian cycle without actually trying all possible sequences of vertices. However, no general method is known. Just to get some idea of the sort of "ad hoc" arguments that are used, let's take a look at two examples. In the graph shown on the left below, there are four vertices each of which is incident to only two edges. Since the cycle must enter and leave each of these vertices, it must include both edges for each such vertex. Thus the cycle must include all the edges shown as jagged lines in the diagram on the right. Now we see that vertex e cannot be connected to b, since b is already included in two edges of the cycle, nor can e connect to d, since this would create a non-Hamiltonian cycle.

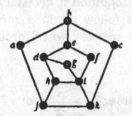

A more abstruse example is shown on the left at the top of page 160. It is formed by taking the graph shown in the center diagram and replacing each of the triangular regions (marked "T") with the subgraph shown on the right. Each of the three triangular subgraphs is oriented such that the edge marked '*' is connected to the vertex at the center of the graph. Note that every vertex is of degree 3, so we can't use the approach that worked on the previous graph. A Hamiltonian cycle in the large graph must enter and leave each triangular subgraph exactly once; furthermore, it must enter each subgraph at one of the three corners of the triangle, follow a path involving each vertex of the subgraph exactly once, and exit via another corner of the triangle.   It turns out (you can satisfy

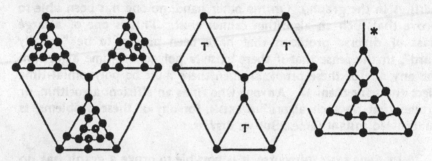

yourself of this by inspection) that any such path must either enter or leave via the edge marked '*'. Hence, all three *-edges must be included in the Hamiltonian cycle. Since these three edges meet at a common point, they cannot all be included in any cycle; hence no Hamiltonian cycle exists.

To convert the memory wheel problem (remember it?) to one of finding a Hamiltonian cycle, we construct a directed graph in which each vertex corresponds to a sequence of length k and two vertices v and w are joined by an edge $v \to w$ if and only if the last $k-1$ digits of the sequence associated with v are the same as the first $k-1$ digits of that associated with w. (This condition is necessary and sufficient for w's sequence being able to follow v's in a memory wheel.) The graphs corresponding to the memory wheel problems for $k = 2$ and $k = 3$ are shown below.

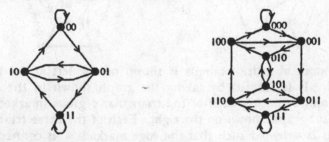

This is all very fine, but it's not particularly useful, since we've already noted that there is no "easy" way to find Hamiltonian cycles. Fortunately, it is also possible to convert the memory wheel problem into that of finding an <u>Eulerian cycle</u>. An Eulerian cycle, as you might guess, is a cycle (not necessarily a simple cycle) that traverses each <u>edge</u> of a graph exactly once. Again, the graph may or may not have directed edges; if it does, each edge is required to be traversed in its designated direction. As we shall see shortly, it is extremely easy to determine whether a graph has an Eulerian cycle and, if it does, to find one.

To convert the memory wheel problem into an Eulerian cycle problem, we construct another directed graph. This time each vertex corresponds to a sequence of length $k-1$ (<u>not</u> k), and two vertices v and w are joined by an edge $v \to w$ if and only if v's sequence is $d_1 d_2 d_3 \ldots d_{k-1}$ and w's is $d_2 d_3 \ldots d_{k-1} d_k$. That is, the last $k-2$ digits of v's sequence must be the same as the first $k-2$ of w's. The edge $v \to w$ corresponds to the sequence $d_1 d_2 d_3 \ldots d_{k-1} d_k$. Thus an edge e can be followed in the cycle by another edge f if and only if f's sequence can follow e's in the memory wheel. (The sequences assigned to the edges have the same sort of "overlap" as did those assigned to the vertices in the Hamiltonian-cycle construction.) Here is the graph corresponding to the memory wheel problem for $k = 3$. The edges' sequences are shown in italics to distinguish them from the vertices'.

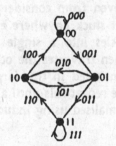

There are two Eulerian cycles in this graph, depending on whether the edges *010* and *101* occur after *001* or after *110*. (Note that $2 = 2^{(2^{3-1}-3)}$.) One cycle is *001, 010, 101, 011, 111, 110, 100, 000*, which gives the memory wheel sequence 00101110.

Euler proved that, in an undirected graph, the following pair of conditions is necessary and sufficient for the existence of an Eulerian cycle (though he probably didn't use that term):

    (*i*)   the graph is connected, and
    (*ii*) every vertex has <u>even</u> degree.

The first condition is obviously necessary. To see that the second is necessary, focus on a particular vertex. Every time we enter it we must then leave it, so if we enter it k times it must be incident to exactly $2k$ edges. The significance of Euler's contribution was that he proved these conditions are not only necessary but also sufficient.

Assume the conditions hold; we wish to prove the existence of an Eulerian cycle. We start at any vertex and wander around the graph until we get stuck, *i.e.*, until we find ourselves at a vertex all of whose edges have already been traversed. Since each vertex has even degree, we cannot get stuck at any vertex other than the one from which we started. At this point we have a cycle (not necessarily Eulerian), as shown on the left below. For instance, we might have started at v, and followed the cycle $vabcdbefawv$. Suppose some edge, such as the jagged edge in the diagram, is not traversed by this cycle. Since the graph is connected, there must be a path connecting this edge to the cycle, as indicated by the dotted lines in the diagram on the right. We then start at the vertex where the path joins the cycle (vertex w in the diagram) and start wandering around the graph some more, using only edges not included in the cycle. Since, even with the cycle edges removed from consideration, every vertex has even degree, we cannot get stuck anywhere except at vertex w. We then combine the two cycles into a single cycle that starts at w, traverses one cycle, and then traverses the other. We look to see if there are any edges not included in this larger cycle, and if so we extend the cycle again. We repeat this until all edges are included in the cycle. (This can be formalised using induction, but we won't take the time to do so here.)

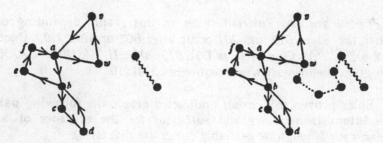

Applying Euler's result to the bridges of Königsberg, we let each land mass be a vertex and each bridge connecting two land masses be an edge, as shown on page 163. There are three vertices with degree 3 and one with degree 5, so there cannot be a cycle that traverses each bridge exactly once. In fact, as we'll see a bit later, it is impossible to find a *path* that traverses each bridge exactly once;

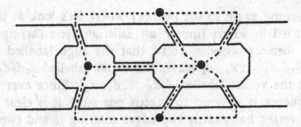

in other words, it is impossible to get from anywhere in Königsberg (or Kaliningrad) to anywhere else in the city, crossing each bridge exactly once.

Now let's consider directed graphs. We'll need to define a few more terms. A directed graph is said to be <u>strongly connected</u> if there is a path from each vertex to every other vertex. (For instance, the sample networks we looked at in chapter 12 are not strongly connected, because there is no path from t to s. There are, of course, paths from s to t, but that's not good enough.) The number of edges entering a vertex (*i.e.*, for a vertex v, the number of directed edges $u \rightarrow v$) is called the <u>in-degree</u> of that vertex, and the number of edges leaving a vertex is called the <u>out-degree</u>. Now we're ready to state the necessary and sufficient conditions for a directed graph having an Eulerian cycle:

(*i*)   the graph is strongly connected, and
(*ii*)   the in-degree of each vertex equals its out-degree.

The proof is similar to that for undirected graphs and is left as an exercise. Condition (*i*) could be replaced by

(*i'*)   the underlying graph is connected,

where the <u>underlying graph</u> of a directed graph is the undirected graph obtained by ignoring the directions assigned to the edges. Clearly condition (*i*) implies condition (*i'*). Since condition (*i'*), together with condition (*ii*), implies the existence of an Eulerian cycle, the two conditions together imply strong connectivity (we can get from any vertex to any other by following the cycle). Hence condition (*i*) is equivalent to condition (*i'*) *assuming condition (ii) holds.*

Returning again to the memory wheel, let's look at the graph we constructed in which finding an Eulerian cycle corresponded to finding a memory wheel. Recall that the edge labelled with the sequence $d_1 d_2 d_3 \ldots d_{k-1} d_k$ leaves the vertex labelled $d_1 d_2 d_3 \ldots d_{k-1}$ and enters the vertex labelled $d_2 d_3 \ldots d_{k-1} d_k$. Since every possible binary sequence is assigned to exactly one edge, it is clear that any particular vertex has exactly two edges entering it and two leaving. Hence each vertex has in-degree = out-degree = 2. Thus we know that there must be an Eulerian cycle, and that therefore a memory wheel of size 2^k must exist for the binary sequences of length k. If we were to formalise the existence proof, we would be able to prove that there are exactly $2^{(2^{k-1}-k)}$ different Eulerian cycles in the graph, and hence the same number of minimum-size memory wheels.

There is an object lesson to be learned from the memory wheel problem, namely that it is quite often possible to transform an easy problem into a hard one. We must always be careful not to jump to the conclusion that, because we have found a hard way to solve a problem, there is no easy way. The memory wheel problem can be solved by solving a Hamiltonian cycle problem, which is hard, but it can also be solved by solving an Eulerian cycle problem, which is easy. In fact, here's an exercise worth pondering: Given some graph G in which we wish to find an Eulerian cycle (if one exists), how could we transform the graph into a second graph G' such that a Hamiltonian cycle in G' corresponds to an Eulerian cycle in G? Using such a transformation, we could solve the Eulerian cycle problem for G by solving instead the Hamiltonian cycle problem for G'. Since the Hamiltonian cycle problem is more difficult than the Eulerian cycle problem, this is obviously not a worthwhile approach, but it is an interesting exercise. Note that, if you could come up with a transformation for the opposite direction, you'd have found an easy way to solve the Hamiltonian cycle problem!

A common variation on the problem of finding a Hamiltonian or Eulerian cycle is that of finding Hamiltonian or Eulerian paths, *i.e.*, paths that go through each vertex or edge exactly once, but that need not end at the vertex from which they begin. The two graphs on pages 159 and 160 that we proved did not have Hamiltonian cycles, *do* have Hamiltonian paths, as shown below. There *are* graphs that lack Hamiltonian paths; the problem of determining whether an arbitrary graph has such a path is just as difficult as determining whether it has a Hamiltonian cycle.

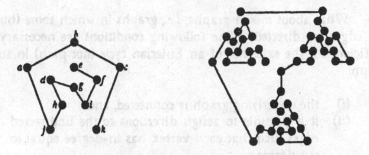

As for determining the existence of Eulerian paths, it is as easy as doing so for cycles. The graph must still be connected, but condition (*ii*) is replaced by the condition that every vertex <u>except two</u> have even degree. The remaining two vertices may be of either even or odd degree. (Note that the sum of the degrees of all the vertices must be even, since it must equal twice the number of edges.) That these modified conditions are necessary and sufficient for the existence of an Eulerian path can be proved either by making a slight modification to the proof for Eulerian cycles, or by using a construction in which the two vertices of odd degree (if they exist) are joined by a new edge, after which an Eulerian cycle must exist (since all vertices are now of even degree). We can then remove the added edge from the cycle and have an Eulerian path. (This proves sufficiency; necessity is, as before, trivial.)

What about finding Eulerian paths in *directed* graphs? We have to be careful, because it is no longer necessary that the graph be strongly connected, so we <u>must</u> use the alternate form for the first condition. We state without proof that the following two conditions are necessary and sufficient for the existence of an Eulerian path in a directed graph:

(*i*) the underlying graph is connected, and

(*ii*) for each vertex except two, the in-degree equals the out-degree; of the remaining two vertices, either both have in-degree equal to the out-degree, or else one has in-degree = out-degree + 1.

Note that the sum of the in-degrees of the vertices must equal the sum of the out-degrees, so if condition (*ii*) is met and one vertex has in-degree = out-degree + 1, there must be exactly one vertex with out-degree = in-degree + 1.

What about mixed graphs, *i.e.*, graphs in which some (but not all) edges are directed? The following conditions are necessary and sufficient for the existence of an Eulerian cycle (not path) in such a graph:

(*i*)   the underlying graph is connected, and
(*ii*) it is possible to assign directions to the undirected
        edges such that each vertex has in-degree equal to
        out-degree.

To see that these conditions are necessary, we observe that if an Eulerian cycle exists, it *tells* us how to direct the undirected edges; we simply orient them in the direction in which they are traversed by the cycle. Hence it is impossible to have an Eulerian cycle without condition (*ii*) being met. To see that the conditions are sufficient, we assume they hold. We then orient the undirected edges as specified in condition (*ii*), after which we are able to find an Eulerian cycle in the directed graph. This cycle is also an Eulerian cycle in the mixed graph.

For example, consider the mixed graph shown on the left on page 167. By directing some of the undirected edges as shown on the right, we can produce a graph in which each vertex to have in-degree = out-degree. (Note that we haven't bothered to assign directions to all of the undirected edges. The remaining undirected edges satisfy the conditions for the existence of an Eulerian cycle in an undirected graph, so it is clearly possible to orient these edges such that the additional in-degree at each vertex is equal to the additional out-degree.) We can then find an Eulerian cycle, such as *abfcbdaedca*.

How might we determine (easily) whether this most recent condition (*ii*) holds? We can use networks! We create two new vertices (the source and sink) and connect them to the graph in the following manner. For each vertex v having in-degree n greater than its out-degree, we add an edge from the source to v with

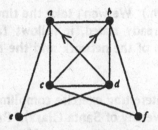 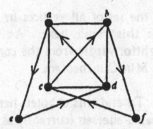

capacity n. For each vertex w having out-degree m greater than its in-degree, we add an edge from w to the sink with capacity m. We then remove all directed edges, and give each undirected edge a capacity of 1. (Note that, in a network flow problem, an undirected edge is equivalent to two directed edges that travel in opposite directions between the same pair of vertices.) In the diagram below, we have performed this transformation on the example above. If we can find a flow in which the edges leaving the source are saturated (and hence the edges entering the sink are also saturated, since their total capacity must, by the construction, equal that of the edges leaving the source), the flow through the undirected edges will tell us

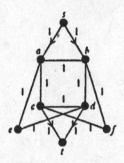

the directions that should be assigned to those edges in order to "balance" the in-degrees and out-degrees at the hitherto unbalanced vertices. By using this construction and the Max Flow Min Cut theorem from chapter 12, we arrive at the following alternative for condition (*ii*):

(*ii'*)   for every subset S of vertices, if the number of
         directed edges leaving vertices in S is n greater
         than the number of directed edges entering vertices
         in S, then the number of undirected edges joining
         S and $V-S$ is at least n.

(*V* is the set of all vertices in the graph.) We won't take the time to prove this result here. As we've already stated, it follows fairly straightforwardly from the construction of the network and the Max Flow Min Cut theorem.

To end this chapter, here's an interesting exercise compliments of Jean Pedersen (currently at the University of Santa Clara). Prove that the graph shown below has no Hamiltonian cycle. Hint: the graph is bipartite!

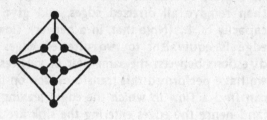

14 Planarity and the Four-Color Theorem

March 16. A graph is said to be <u>planar</u> if it can be drawn in the plane with no crossing edges. A classic problem based on planarity is the "utility problem". Suppose there are three people—Jack, Jill, and Judy—living in separate houses, and also three utilities—water, gas, and electricity—each supplied by a different plant. We wish to connect each of the three houses to each of the three plants, but we don't want any of the nine connections to cross each other. (Perhaps all of the utilities are supplied by cables or pipes buried just beneath the surface, and if two were to cross we might damage one conduit while installing the other. Oh well, nobody ever said this problem was practical, just that it was a classic.) We can easily make eight of the connections, but we run into trouble with the ninth, as shown below (left). Struggle as we may, we will be unable to make all nine connections. (We could cheat and run a conduit underneath one of the buildings, but this is not considered valid.) This graph is <u>non-planar</u>. The graph is shown on the right below; it is the <u>complete bipartite graph</u> on two sets of three vertices, meaning that it contains two sets of three vertices along with <u>all</u> edges joining a vertex in one set with one in the other set. This graph is usually denoted by $K_{3,3}$.

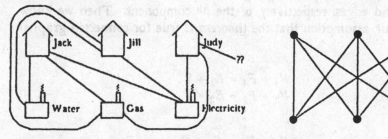

Planar graphs have many interesting properties. Let's look at some planar graphs and see what we can observe. The graphs shown on the following page are planar "projections" of a cube and tetrahedron. (They are what we would see if we built the outlines of those figures (using wires for the edges, say) and then looked at them from points very close to the center of one face.) We'll use V to represent the number of vertices in a graph, E the number of edges, and F the number of faces, where a <u>face</u> is a region that is bounded by edges of the graph and contains no other edges of the graph.

G. Pólya et al., *Notes on Introductory Combinatorics*, Modern Birkhäuser Classics,
DOI 10.1007/978-0-8176-4953-1_14, © Birkhäuser Boston, a part of Springer
Science+Business Media, LLC 2010

(For example, the graph of the cube has a square face in the center, four trapezoidal faces surrounding the square face, and one square "exterior" face, bounded by the four outer edges and consisting of the infinite region "outside" the graph.) The cube has $F = 6$, $V = 8$, and $E = 12$. The tetrahedron has $F = 4$, $V = 4$, and $E = 6$. In both cases, we observe that $V + F = E + 2$. This equation is often called <u>Euler's formula</u>, and is asserted by the following theorem: Any connected planar graph has $V + F = E + 2$. A planar graph that is not connected satisfies $V + F = E + 1 + C$, where C is the number of connected components. (See page 184 for a formal definition of "connected components".)

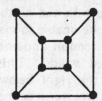 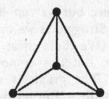

We'll first prove the theorem for disconnected graphs by using the connected case (which we'll prove later). Consider a planar graph with $C \geq 2$. Let V_i, F_i, and E_i be the number of vertices, faces, and edges, respectively, of the ith component. Then we know (from our assumption that the theorem is true for connected graphs) that

$$V_1 + F_1 = E_1 + 2$$
$$V_2 + F_2 = E_2 + 2$$
$$\vdots$$
$$V_C + F_C = E_C + 2$$

and hence

$$(V_1+V_2+ \cdots +V_C) + (F_1+F_2+ \cdots +F_C) = (E_1+E_2+ \cdots +E_C) + 2C.$$

Clearly, $(V_1+V_2+ \cdots +V_C) = V$ and $(E_1+E_2+ \cdots +E_C) = E$. When we sum the F_i, however, we count each interior face exactly once, but we count the exterior face C times, once per component. (The exterior face of a disconnected graph is an infinite region with two or more "holes" in it, one per component of the graph.) Since F only counts this region once, we find that $(F_1+F_2+ \cdots +F_C) = F + (C-1)$.

Thus

$$V + F + (C-1) = E + 2C,$$

and the theorem follows.

Now we shall prove the theorem for connected graphs by induction on the number of edges. We start the induction by considering a graph that has one vertex and no edges. (We cannot have more than one vertex and still have no edges, since the graph is supposed to be connected.) This graph has one face, namely the exterior face. So $V = 1$, $F = 1$, and $E = 0$. Since $1 + 1 = 0 + 2$, the theorem holds. Now we perform the induction by seeing what happens when an edge is added to the graph. There are two cases: (1) the edge could lead to a vertex not previously included in the graph, or (2) the edge could join two old vertices. (The new edge cannot join two vertices neither of which was yet in the graph, because this would result in a disconnected graph.)

In case (1), the new edge (the dotted edge in the diagram on the left below) does not create any new faces. (The new edge becomes part of an existing face, the border of which includes <u>both</u> "sides" of the edge.) So if the graph prior to the addition of the dotted edge had V vertices, F faces, and E edges, then the graph including the dotted edge has $V' = V + 1$ vertices, $F' = F$ faces, and $E' = E + 1$ edges. From our induction hypothesis, we know that $V + F = E + 2$; hence we conclude that $V + 1 + F = E + 1 + 2$, and thus $V' + F' = E' + 2$. In case (2), the new edge divides an existing face into exactly two faces, as shown in the center diagram below. (The new edge cannot divide more than one face without crossing another edge, as in the diagram on the right, and this is forbidden since the graph is planar.) Thus $V' = V$, $F' = F + 1$, and $E' = E + 1$, and we again find that the induction hypothesis ($V + F = E + 2$) implies $V' + F' = E' + 2$. The theorem is proved.

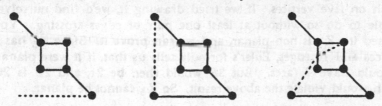

Note that, if we consider the boundaries of *all* the faces of a graph, each edge appears twice, once for each "side" of the edge. In general each edge will appear in two different faces, but there are exceptions. For example, consider the graph shown below. This graph has three faces: there are two triangular faces—*abca* and *defd*—and an exterior face—*acdefdcba*—which traverses the edge *cd*

in both directions. The general rule still holds:

$$F_1 + 2F_2 + 3F_3 + 4F_4 + 5F_5 + \cdots = 2E,$$

where F_i is the number of faces with exactly i edges. Let's assume there are no faces with fewer than three edges. This is equivalent to saying that (1) there are no "self-loops" (edges joining a vertex to itself, as shown on the left below), since this would result in a face bounded by a single edge, (2) no two vertices are joined by more than one edge (as shown on the right below), since this would result

in a two-edged face, and (3) the graph does not consist of a single edge, since the exterior face would then be bounded by the two sides of this edge. *Under these restrictions*, we can conclude that

$$3F = 3F_3 + 3F_4 + 3F_5 + \cdots$$
$$\leq 3F_3 + 4F_4 + 5F_5 + \cdots = 2E.$$

What good is this result? Well, let's look at K_5, the complete graph on five vertices. If we tried drawing it, we'd find ourselves unable to do so without at least one pair of edges crossing. You guessed it—K_5 is non-planar, and we can prove it. Since K_5 has 5 vertices and 10 edges, Euler's formula tells us that, if it were planar, it would have 7 faces. But $3F$ would then be 21, and $2E$ is 20, which would violate the above result. So K_5 cannot be planar.

If we try applying the same proof to $K_{3,3}$, we have less luck. Since $V = 6$ and $E = 9$ for this graph, Euler's formula tells us that a planar representation of $K_{3,3}$ would have to have 5 faces. Since $3 \cdot 5 \leq 2 \cdot 9$, we haven't proved anything. However, we can observe that, since the graph is bipartite, it contains no triangles. Hence every face must include four or more edges. When $F_3 = 0$, we find that

$$4F = 4F_4 + 4F_5 + 4F_6 + \cdots$$
$$\leq 4F_4 + 5F_5 + 6F_6 + \cdots = 2E.$$

Since $4F = 20$ and $2E = 18$, $K_{3,3}$ cannot be planar.

We now come to one of the most important theorems dealing with planarity. It is important not so much because it has useful applications, but rather because it was the earliest non-topological characterisation of planar graphs. The theorem is by Kuratowski and is, appropriately enough, called <u>Kuratowski's Theorem</u>. We observe that any graph that contains K_5 or $K_{3,3}$ as a subgraph must be non-planar. Furthermore, placing additional vertices upon the edges of one of these graphs cannot make it planar (assuming no vertex joins two hitherto disjoint edges). For instance, the graphs shown below are non-planar, and so are any graphs that contain them as subgraphs. A graph that is isomorphic to some graph G, aside from such additional vertices along edges, is said to be a <u>generalised</u> graph of G. (Two graphs G and G' are <u>isomorphic</u> if there is a one-to-one correspondence of the vertices such that two vertices in G' are joined by an edge if and only if the corresponding vertices in G are.)

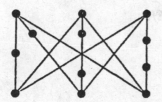 

Kuratowski's Theorem asserts that *every* non-planar graph must contain either a generalised K_5 or a generalised $K_{3,3}$ as a subgraph. Generalised K_5's and $K_{3,3}$'s are often referred to as <u>Kuratowski subgraphs</u>. For example, consider the graph shown on the left on the following page, which is known as "Petersen's graph".

This graph is non-planar, and so it must contain a Kuratowski subgraph. Indeed it does, as indicated in the diagram on the right. The graph is a generalised $K_{3,3}$; the two sets of vertices forming the vertices of the bipartite graph are shown as dark and light circles, and the nine paths forming the "edges" of the generalised graph are drawn solid (the remaining edges being dotted).

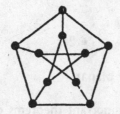 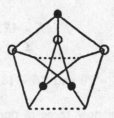

Unfortunately, the proof of Kuratowski's Theorem is much too involved for us to attempt to present in these notes. If you're interested, you can find proofs in [Liu], pp. 212-220, and [Harary], pp. 109-112. The theorem is mathematically elegant but, as we have already noted, it is not very useful in practice. The most efficient algorithms for testing whether graphs are planar involve actually embedding the edges of the graphs in the plane. When such an algorithm determines that a given graph is non-planar, it is not necessarily clear how to go about isolating a Kuratowski subgraph within that graph, although one is known to exist.

Returning to some of our earlier results, we can manipulate the formulas involving V, F, and E to come up with some additional theorems. For example, since $V + F = E + 2$ and $3F \leq 2E$, we can conclude that

$$E + 2 \leq V + 2E/3$$

and hence

$$E \leq 3V - 6.$$

This means that, given the restriction against one- and two-sided faces, no planar graph can have more than $3V - 6$ edges. Since, among V vertices, it is possible to have $\binom{V}{2} = (V^2 - V)/2$ edges, this can be interpreted as saying that planar graphs must have relatively few edges compared to most graphs. Note that K_5 has $E > 3V - 6$, so we have again verified its non-planarity. If we know there are no triangular faces, the preceding result becomes $E \leq 2V - 4$, which verifies the non-planarity of $K_{3,3}$.

We now look at a concept known as the <u>dual</u> of a planar graph. The dual is a graph that consists of one vertex for each face of the original graph, with edges connecting two vertices in the dual if and only if the corresponding faces shared an edge in the original graph. The diagram below shows the graph of the cube (dark circles and solid lines) together with its dual (white circles and dotted lines). Note that the dual is always itself planar, and that the dual of the dual is the original graph. We can establish a one-to-one correspondence between features of the two graphs; each face in the original graph corresponds to a unique vertex in the dual, and vice versa, and each edge in the dual crosses a unique edge in the

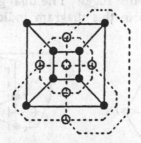

original. Thus, if a graph has V vertices, F faces, and E edges, and its dual has V' vertices, F' faces, and E' edges, then we might assert that $V = F'$, $F = V'$, and $E = E'$. We may therefore rephrase any of our earlier results relating V, F, and E by applying them to the dual and using this correspondence. For example, since the dual must have $E' \leq 3V' - 6$, we can assert that any planar graph must have $E \leq 3F - 6$. We must not be too hasty, however. Recall that we have been assuming there are no faces with fewer than three edges in our planar graphs. If we apply these restrictions to the dual, what does it imply regarding the original graph? A face in the dual corresponds to a vertex in the original graph. If the vertex in the original graph has degree k, it means there are k edges incident to that vertex. Each of these k edges is crossed by an edge in the dual, resulting in a k-sided face. For instance, the vertex shown on the left on the following page has degree five; thus five faces of the original graph meet at the vertex. Each of these faces corresponds to a dual vertex, as shown in the center diagram. These dual vertices are then connected by dual edges as shown on the right to form a dual face with five edges.

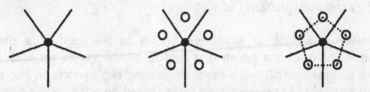

   Thus, for the dual graph to have no faces with fewer than three edges, the original graph must have no vertices of degree two or less.   Interestingly, there are graphs satisfying this condition yet having duals containing doubled edges or self-loops.   For instance, consider the graph shown on the left below.   The face *abehgda* shares two edges (*ab* and *gh*) with the exterior face.   Thus the dual graph has a doubled edge.   But the dual graph does *not* have any faces with fewer than three edges!   The dual graph is shown on the right; the dotted edges in this diagram indicate the edges of the graph on the left.

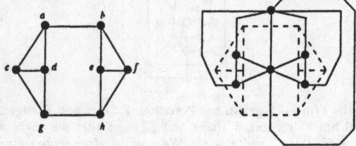

   If we restrict ourselves to planar graphs in which all vertices are of degree three or greater, we can derive some additional results. Since, in the dual graph, we know that $E' \leq 3V' - 6$, we know that the original graph has $E \leq 3F - 6$. Hence

$$6F_3 + 6F_4 + 6F_5 + 6F_6 + 6F_7 + \cdots = 6F \geq 2E + 12.$$

If we also continue to assume that the graph has no self-loops or doubled edges, then we know that there are no faces with fewer than three edges, so

$$3F_3 + 4F_4 + 5F_5 + 6F_6 + 7F_7 + \cdots = 2E.$$

Subtracting this equation from the preceding inequality, we find that

$$3F_3 + 2F_4 + F_5 - F_7 - 2F_8 - 3F_9 - \cdots \geq 12,$$

which can be rewritten as

$$3F_3 + 2F_4 + F_5 \geq 12 + F_7 + 2F_8 + 3F_9 + \cdots \geq 12.$$

This implies that any planar graph that (1) has no self-loops or doubled edges and (2) has no vertex with degree less than three, must have at least four "small" faces, where a small face is defined as being a triangle, quadrilateral, or pentagon. If it has no triangles or quadrilaterals it must have at least 12 pentagonal faces.

In solid geometry, a regular polyhedron is a solid of which all the faces are regular polygons and are congruent to one another. Each vertex of a regular polyhedron is incident to the same number of edges. It was well-known even to the ancient Greeks that there are exactly five regular polyhedra. We can prove this using planar graphs. We first observe that we can convert any polyhedron into a planar graph by painting the vertices and edges of the polyhedron on some solid, flexible object, such as an inflated balloon, and then cutting a hole inside one face and flattening the balloon onto a plane surface. (We would probably want to deflate the balloon before cutting the hole; it results in less noise.) If the polyhedron is regular, then in the resulting graph each vertex will have the same degree and each face will have the same number of edges. How many such graphs are there? We note that the vertices must be of degree three or more, since it requires the intersection of three planes (each face of the polyhedron specifies a plane) to define a point (a vertex of the polyhedron), and the intersection of each pair of planes defines an edge entering that vertex. Hence we know the graph must include some small faces, and since all faces are identical there are three cases: (1) all the faces are triangles, (2) all the faces are squares, or (3) all the faces are pentagons. Furthermore, since all vertices are of the same degree, they cannot be of degree greater than five. (If there are V vertices all of degree k, then there are kV "endpoints of edges" and hence $kV/2$ edges. If $k \geq 6$, this violates the restriction that E be no greater than $3V - 6$.) We've already noted that the vertices cannot be of degree less than three, so the vertices are of degree three, four, or five. We have therefore nine cases.

If all the faces are triangular and the vertices are of degree three, we get the tetrahedron. The octahedron, which we have encountered before, has eight triangular faces and six vertices of degree four. The icosahedron has twenty triangular faces and twelve vertices of degree five. Moving on to square faces, the hexahedron (more commonly referred to as a cube) has six such faces and eight vertices of degree three. When there are no triangular faces, we know that $E \leq 2V - 4$ (see page 175); hence we cannot let all the vertices be of degree greater than three. There is therefore no other regular polyhedron with square faces. Similarly, there is only one regular polyhedron with pentagonal faces—the dodecahedron, which has twelve such faces and twenty vertices of degree three. (We saw the graph of the dodecahedron in chapter 13.) According to our results regarding planar graphs, these are the only regular polyhedra possible. We observed on the midterm that the cube and the octahedron are each other's duals. The dodecahedron and the icosahedron are also duals. The tetrahedron is its own dual.

We conclude this chapter by examining a problem that was recently solved after withstanding the efforts of mathematicians for over a century, to wit, finding a proof of the Four-Color Theorem. This theorem states that the faces of any planar graph can be colored using four colors such that no two adjacent faces have the same color. Two faces are considered to be adjacent only if they have an edge in common; two faces having one or more vertices in common without sharing an edge may be given the same color. The diagram below shows one way of coloring the dodecahedral graph; each face has been marked with a number from 1 to 4 indicating the color assigned to that face.

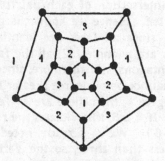

It appears that the four-color problem originated with Francis Guthrie in 1852. In 1879, Kempe thought he had found a proof, but in 1890 Heawood discovered a flaw in Kempe's work. Heawood was able to show that any planar graph can be colored with *five* colors. Over the years several mathematicians have thought they had proved the theorem, but it was only recently (1976) that a proof was found that, at least to date, appears flawless. The proof was constructed by Haken and Appel at the University of Illinois and involved using 1200 hours of computer time to perform an extremely complex case analysis, resulting in a set of 1936 graphs. Haken and Appel claim that every planar graph (aside from some trivial cases, such as the null graph, for which the theorem is clearly true) must contain one of the 1936 graphs as a subgraph, and that each of these 1936 possible subgraphs is "reducible". A "reducible" subgraph R is a subgraph with the property that, for any graph G containing R, one can produce a smaller graph G' such that four-coloring G' shows how to four-color G. If any graph exists that cannot be colored using four colors, there must be a smallest such graph G; since it must contain a reducible subgraph, we can construct a smaller graph G' that, being smaller than G, must be four-colorable. By the definition of a reducible subgraph, we must therefore be able to four-color G, contradicting our original assumption. (At one point there was a rumor that a 1937th case (fortunately also reducible) had been found, but to our knowledge this has never been confirmed. If true, it would cast a shadow on the whole proof.) People are still searching for simpler proofs of the theorem; until one is found, it is obviously impractical for us to present a proof of the Four-Color Theorem. We *can*, however, prove the Five-Color Theorem, and we shall now do so.

We start by assuming that all vertices are of degree three. If there is a vertex of degree two, as in the leftmost graph on the following page, we can eliminate that vertex and combine its two edges into a single edge, as shown in the second graph. This obviously does not affect the colorability of the graph. (You are encouraged to examine the diagrams to convince yourself of this.) If there is a vertex of degree k greater than three, as in the third graph, we can replace it with a k-sided polygon as shown in the rightmost diagram. If the new graph is five-colorable, then we can simply ignore the color of the newly created face and thereby obtain a coloring for the original graph. Hence, if we can prove the Five-Color Theorem for graphs in which all vertices have degree three, we will have proved it in general.

The proof is by induction on the number of faces. Clearly, if there are fewer than six faces, five colors must be sufficient. We shall now show that, if it is possible to five-color all graphs with n faces, it is possible to five-color all graphs with $n+1$ faces. Consider any graph with $n+1$ faces. Since we are assuming all vertices are of degree three, we know that there must be a small face. Suppose it is a triangle; the graph thus contains a subgraph of the form shown on the left below. We then replace this portion of the graph by the subgraph shown in the center diagram, condensing the triangle to a single vertex. By the induction hypothesis we must be able to five-color this new graph, since it has one less face than the one we started with. Suppose, in this coloring, the three faces in the center diagram are assigned the colors x, y, and z as shown. We can then color the original graph as shown on the right, where w is some color other than x, y, and z. (We know there are two such colors.)

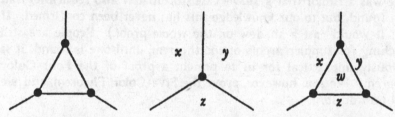

Meanwhile, suppose there are no triangles, but suppose there is a quadrilateral face, as shown on the left at the top of the following page. We have labelled the faces for reference. Suppose x is adjacent to x', as suggested by the dotted lines in the center diagram. If this is the case, then y cannot be adjacent to y'. Similarly, if y is adjacent to y', then x cannot be adjacent to x'. The two cases are equivalent; we shall assume x and x' are adjacent. (If neither pair is adjacent the proof is even simpler, and is left as an exercise.) We remove two edges and four vertices so as to combine y, z, and y' into a single region w, as shown on the right. The induction hypothesis tells us we can five-color this new graph. We can then color the original graph by letting both y and y' be colored using the color applied to w, and coloring z with some color other than those used for x, x', and w.

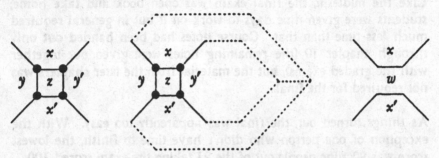

So far everything we've done would apply equally well to a proof of the Four-Color Theorem. (In fact, what we have shown is that triangular and square faces are reducible subgraphs.) Kempe thought he had handled the pentagonal case as well, but Heawood proved him wrong. Five colors, however, are enough to handle this case. Suppose there are no triangles or quadrilaterals; there must be a pentagonal face, as shown on the left below. We again assume that some two of the five regions surrounding z, though not sharing a vertex of the pentagon, are mutually adjacent; if this is not the case the proof is simpler. Suppose x and x' are adjacent; this implies that y and y' cannot be adjacent. We therefore remove two edges and four of the five vertices as shown in the diagram on the right. We know we can five-color this smaller graph. We then color the original graph by letting y and y' be colored using the color assigned to w, and using the fifth color (the one not used on any of x, x', q, or w) to color z. This concludes the proof of the Five-Color Theorem.

If you want to read more about the Four-Color Problem, [Saaty-Kainen] includes just about everything you might want to know, including a brief history of the problem, the theory underlying the recent Haken-Appel proof, and a description of the various reformulations of the problem which have been developed over the hundred-odd years of its existence.

15 | Final Examination

Like the midterm, the final exam was open book and take home; students were given nine days to work on it but in general required much less time than that. Course notes had been handed out only through chapter 10 (the remaining notes were given out together with the graded exams), but the material from the later chapters was not required for the final.

As things turned out, the final was apparently too easy. With the exception of one person who didn't have time to finish, the lowest score was 90; nine people out of the 21 taking the exam scored 100.

Problem 1 (20 points).

This problem doesn't directly involve anything covered in the course, but is instead an easy problem from graph theory. It is included on this exam because it demonstrates the sort of constructions that arise in proving things about graphs and, as you have no doubt noticed, graphs play a significant role in many areas of combinatorics.

In chapter 6 (pages 79 and 80) we defined what we meant by a graph, and what it meant for a graph to be <u>connected</u>. We now define the <u>complement</u> of a graph. Given a graph G, consisting of a set of vertices V and a set of edges E, we define the complement $\overline{G}$ of G to be the graph with the same set V of vertices, but with edges $\overline{E}$ such that an edge is included in $\overline{E}$ if and only if it is *not* included in E. The two graphs shown below are complements of each other.

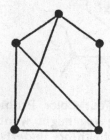 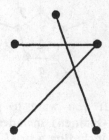

Notice that the graph on the left is connected, but the one on the

G. Pólya et al., *Notes on Introductory Combinatorics*, Modern Birkhäuser Classics, DOI 10.1007/978-0-8176-4953-1_15, © Birkhäuser Boston, a part of Springer Science+Business Media, LLC 2010

right is not. Prove that, if a graph G is not connected, then $\overline{G}$ is. Is the converse true; that is, if G is connected, must $\overline{G}$ be disconnected?

Problem 2 (30 points).

Consider the complete graph on 17 vertices (K_{17}). Prove that, if the edges of K_{17} are colored using <u>three</u> colors, then there must be at least one monochromatic triangle. Is this sufficient to prove that $\Re(3,3,3,2) = 17$? Explain.

Problem 3 (30 points).

Given a collection of sets $S_1, S_2, S_3, \ldots, S_n$, not necessarily disjoint, we define a <u>system of distinct representatives</u> (SDR) to be a set of n distinct elements $\{x_1, x_2, x_3, \ldots, x_n\}$ such that x_k is in S_k for each k from 1 to n. The ordering of the elements x_1 through x_n <u>is</u> considered significant. For instance, if $n = 3$ and the three sets are $\{a,b\}$, $\{a,c\}$, and $\{b,c\}$, then there are 2 different SDRs: $\{a,c,b\}$ and $\{b,a,c\}$.

Suppose the sets $S_1, S_2, \ldots, S_n$ contain respectively $2, 3, \ldots, n+1$ elements. Show that there must be at least 2^n different SDRs. Exhibit such sets in which 2^n is the exact number of SDRs.

Problem 4 (20 points).

Here's one for old times' sake. Find the generating function

$$\sum_{k=0}^{\infty} E_k x^k = E_0 + E_1 x + E_2 x^2 + \cdots$$

in which E_n is the number of ways of changing n cents using pennies, nickels, dimes, quarters, and half dollars, using <u>at least one</u> of each type of coin.

SOLUTIONS

Problem 1 (20 points).

A lot of what is about to follow could be omitted if we had formally established what is meant by certain terms in graph theory, such as "connected component". (We used this term in an informal sense in chapter 14.) Students who knew enough graph theory to be able to omit definitions of such terms were permitted to. For completeness in this solution, however, we'll start with what we already know.

If G is not connected, it means (by definition) that there are some two vertices, say s and t, such that there is no path in G from s to t. Let S be the set of vertices that do have paths between themselves and s. (The set S, together with all the edges between pairs of vertices in S, constitutes a connected component of G.) Let T be the set of all vertices in V that are not contained in S. (T may consist of more than one connected component; this doesn't make any difference to our proof.) Note that s is in S and t is in T, so neither set is empty. Note also that, for any vertices u in S and v in T, the graph G cannot include the edge uv, since it would mean that there was a path from s to v via u, and hence v would be in the set S. Thus, for all vertices u in S and v in T, the complement graph $\overline{G}$ must include the edge uv.

We wish to show that $\overline{G}$ is connected. To do this we need only show that, for any two vertices x and y in $\overline{G}$, there is a path from x to y. If x is in S and y is in T, or vice versa, then they are directly joined by an edge in $\overline{G}$, and we are through. If both x and y are in S, then we pick any vertex in T (recall there must be at least one such vertex), say t. $\overline{G}$ includes the edges xt and ty, so x and y are joined by a path involving two edges. Similarly, if both x and y are in T, they are joined by a path of two edges going through the vertex s. Hence every pair of vertices in $\overline{G}$ is connected by a path, and thus by definition $\overline{G}$ is connected.

In fact, we have shown that if G is not connected it implies that every pair of vertices in $\overline{G}$ is connected by a path of one or two edges. Thus, if G is connected but contains two vertices that are not joined by any path of fewer than three edges, G cannot be the complement of a disconnected graph, and therefore $\overline{G}$ must also be connected. The following pair of graphs are complementary, and are both connected. (In fact, this is an example of a graph that is isomorphic to its own complement.)

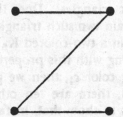 

Notice that these graphs contain pairs of vertices that are three edges apart. It is somewhat trickier coming up with graphs in which all pairs of vertices are within two edges of each other, but that have connected complements. A single vertex with no edges qualifies vacuously as such a graph; the pentagon also works. (The pentagon is another graph that is isomorphic to its complement.) There are in fact an arbitrary number of such graphs; two examples are shown below.

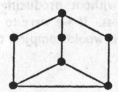 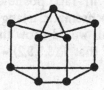

Problem 2 (30 points).

Consider an arbitrary vertex of K_{17}; call it v. Since the graph is complete, v must have edges joining it to each of the remaining 16 vertices. At least 6 of these edges must be colored using the same color since, if each of the three colors is used no more than 5 times, this accounts for at most $5+5+5 = 15 < 16$ edges. So one color, call it c_1, must be used on at least 6 of these 16 edges. Let $u_1, u_2, \ldots, u_6$ be the vertices at the "other ends" of these six edges. Since the graph is complete, all pairs of vertices have edges joining them, so the six vertices u_1 through u_6 form a K_6. If any edge of this K_6, say $u_i u_j$, is assigned color c_1, then there is a c_1 triangle: $v u_i u_j$.

Otherwise, all of the edges of the K_6 must be colored using the remaining two colors, and we already know that this must result in a monochromatic triangle. Our proof is thus complete.

As an aside, recall that we actually showed that a two-colored K_6 must contain at least <u>two</u> monochromatic triangles. Does this imply that a three-colored K_{17} must also contain two such triangles? Not quite. It is possible for the two triangles in a two-colored K_6 to share a common edge (there is a unique coloring with this property); if this common edge is instead colored using color c_1, then we get only a single triangle. On the other hand, there are ten other vertices to contend with, and it's not difficult to show that, in fact, there must be at least two monochromatic triangles altogether. The actual minimum number of such triangles in a three-colored K_{17} is, to the best of our knowledge, unknown.

Have we proven that $\Re(3,3,3,2) = 17$? The Ramsey number is defined to be the *minimum* number such that, if $N \geq \Re(3,3,3,2)$ and the edges of K_N are colored using three colors, there must be a monochromatic triangle. The preceding proof is not sufficient to establish that 17 is the minimum; all we can assert is $\Re(3,3,3,2) \leq 17$. It is in fact possible to three-color a K_{16} without producing a monochromatic triangle, but it's a tedious process. If we were to do so, this would show that $\Re(3,3,3,2) > 16$, and this would complete the proof that $\Re(3,3,3,2) = 17$.

Problem 3 (30 points).

Some people approached this problem by first asserting something to the effect of, "The minimum number of distinct SDRs is achieved when there is maximum containment among the sets, *i.e.*, when $S_1 \subset S_2 \subset S_3 \subset \cdots \subset S_n$." These people then proceeded to show that, in this situation, there are exactly 2^n distinct SDRs, and claimed that therefore there are always at least that many. This is going about it backwards. How do we <u>know</u> that the minimum occurs when there is maximum containment? It may be intuitively obvious, but that's not a proof. In fact, the most reasonable way to prove it is to show that maximum containment results in 2^n SDRs and that there are always at least 2^n SDRs, and *conclude* that the minimum occurs when there is maximum containment. Just to *claim* that the minimum occurs with maximum containment, and use this claim to conclude that there are always at least 2^n SDRs, is circular logic. So let's prove this the right way!

If $n = 1$, then S_1 consists of two elements, x_1 and x_2, and there are obviously $2^1 = 2$ possible SDRs, namely $\{x_1\}$ and $\{x_2\}$. So the result is true for $n = 1$. We'll use induction to prove it is true for any finite n.

Suppose the result is true for $n-1$. That is, given $n-1$ sets containing respectively $2, 3, \ldots, n$ elements, there are at least 2^{n-1} different SDRs. Now consider the situation when another set, with $n+1$ elements, is added to the collection. We know, by the induction hypothesis, that there are at least 2^{n-1} ways to choose $n-1$ distinct elements, one from each of the first $n-1$ sets (where the order of selection is considered to be significant). For each such choice, there are at least two choices for the element to represent the last set. Why? The set contains $n+1$ elements, and at most $n-1$ of them can be included among the elements already in the SDR. Therefore there are at least 2 elements that have not yet been chosen and are therefore eligible to represent this set. Since there are at least 2^{n-1} ways to choose the first $n-1$ elements and, for each choice, at least 2 ways to choose the last element, there are at least $2 \cdot 2^{n-1} = 2^n$ ways to choose the n elements of the SDR.

If $S_1 \subset S_2 \subset S_3 \subset \cdots \subset S_n$, then there are again exactly 2 SDRs for the single set S_1. This time, however, when we add the nth set to the collection, there are, for each possible SDR for the first $n-1$ sets, exactly two choices for the element that will represent S_n. This is because, no matter which elements are chosen to represent the first $n-1$ sets, these $n-1$ elements must be contained in S_n, and thus exactly $n-1$ of the $n+1$ elements in S_n are ineligible to represent it, leaving exactly two choices. Thus, by induction, there are exactly 2^n SDRs for these sets. One specific example of such a case is the collection of sets $\{a,b\}$, $\{a,b,c\}$, $\{a,b,c,d\}$. There are exactly eight distinct SDRs for these sets, namely $\{a,b,c\}$, $\{a,b,d\}$, $\{a,c,b\}$, $\{a,c,d\}$, $\{b,a,c\}$, $\{b,a,d\}$, $\{b,c,a\}$, and $\{b,c,d\}$.

Problem 4 (20 points).

The only difficulty with this problem seemed to be some confusion over just what constitutes a "generating function". The summation

$$\sum_{k=0}^{\infty} E_k x^k = E_0 + E_1 x + E_2 x^2 + \cdots$$

is a generating function for the sequence E_0, E_1, E_2, etc., but is not very useful in this form. For instance, in chapter 3, we found a generating function that could be written in the above form, where E_n was the number of ways of changing n cents using five types of coins (without the restriction that each type be used at least once). However, writing the function as a summation like this is not useful, since it does not help us find the coefficients E_n. On the other hand, the set of recursion formulas we proceeded to find for computing E_n were *not* themselves a generating function; they were simply a means for computing the *coefficients* of the generating function. The function itself was

$$\frac{1}{(1-x)(1-x^5)(1-x^{10})(1-x^{25})(1-x^{50})}.$$

The whole <u>idea</u> of using a generating function is that we want to be able to deal with the infinite sequence as a single unit; we want a <u>finite</u> form that "embodies" the infinite sequence. The infinite summation tells us what the generating function "is" in the sense that it tells us what it represents. The finite formula shown above tells us what the generating function "is" in the sense that it gives us a mathematical means for expressing it. Since the infinite summation is trivial to specify (and in fact was given as part of the statement of the problem), it is obviously the finite formula that is called for.

There are two "easy" ways to find the desired function—using either the method or the result from chapter 3. Using the result is easier, so let's do it that way first. We observe that one way to provide change for n cents, using at least one of each type of coin, is to start by setting aside the five required coins, after which we may use any combination of coins for the remaining amount. The five required coins have a value of $1+5+10+25+50 = 91$ cents. Thus, to change n cents, we set aside 91 cents and then change $n-91$ cents using any coins we wish. Let E'_n represent the number of ways of changing n cents, <u>without</u> the restriction that at least one of each type of coin be used. We know from chapter 3 that

$$\sum_{k=0}^{\infty} E'_k x^k = \frac{1}{(1-x)(1-x^5)(1-x^{10})(1-x^{25})(1-x^{50})}.$$

Meanwhile, we have just shown that $E_n = E'_{n-91}$. Hence

$$\sum_{k=0}^{\infty} E_k x^k = \sum_{k=0}^{\infty} E'_{k-91} x^k$$

$$= \sum_{k=91}^{\infty} E'_{k-91} x^k$$

(since we know $E'_n = 0$ for all $n < 0$). We then let $j = k-91$ on the righthand side and find

$$\sum_{k=0}^{\infty} E_k x^k = \sum_{j=0}^{\infty} E'_j x^{j+91}$$

$$= x^{91} \cdot \sum_{j=0}^{\infty} E'_j x^j$$

$$= x^{91} \cdot \frac{1}{(1-x)(1-x^5)(1-x^{10})(1-x^{25})(1-x^{50})}$$

$$= \frac{x^{91}}{(1-x)(1-x^5)(1-x^{10})(1-x^{25})(1-x^{50})},$$

which is the desired result.

Meanwhile, we could also have started from scratch and used the method described in chapter 3. We would find that we could use one penny, or two, or three, etc., *but not zero*, and would write these choices as the infinite sum, $x + x^2 + x^3 + \cdots$. Similarly, we could have one nickel, or two, or three, etc., giving us the sum $x^5 + x^{10} + x^{15} + \cdots$. Continuing in this fashion, we eventually take the product of these sums, thus:

$$(x + x^2 + x^3 + \cdots)$$
$$\cdot (x^5 + x^{10} + x^{15} + \cdots)$$
$$\cdot (x^{10} + x^{20} + x^{30} + \cdots)$$
$$\cdot (x^{25} + x^{50} + x^{75} + \cdots)$$
$$\cdot (x^{50} + x^{100} + x^{150} + \cdots)$$

Each term of the product with an exponent of n corresponds to one way of selecting a term from each sum such that the exponents sum to n. Now, since

$$x + x^2 + x^3 + \cdots = x \cdot (1 + x + x^2 + \cdots) = x \cdot \frac{1}{1-x}$$

(and similarly for the other sums), we would find that the desired generating function is

$$\frac{x}{1-x} \cdot \frac{x^5}{1-x^5} \cdot \frac{x^{10}}{1-x^{10}} \cdot \frac{x^{25}}{1-x^{25}} \cdot \frac{x^{50}}{1-x^{50}},$$

which is equivalent to our earlier answer.

16 Bibliography

Alexandru T. Balaban (editor), *Chemical Applications of Graph Theory.*

Claude Berge, *Principes de Combinatoire.*

Lester Ford, Jr., and Delbert Fulkerson, *Flows in Networks.*

Marshall Hall, Jr., *Combinatorial Theory.*

Frank Harary, *Graph Theory.*

Frank Harary (editor), *A Seminar on Graph Theory.*

Donald E. Knuth, *The Art of Computer Programming.*

C. L. Liu, *Introduction to Combinatorial Mathematics.*

E. Netto, *Lehrbuch de Combinatorik* (2nd edition, 1927).

George Pólya, *How to Solve It.* [HSI]

George Pólya, *Mathematical Discovery.* [MD]

George Pólya, *Mathematics and Plausible Reasoning.* [MPR]

George Pólya and Gabor Szegö, *Problems and Theorems in Analysis.* [P-Sz]

John Riordan, *An Introduction to Combinatorial Analysis.*

Herbert J. Ryser, *Combinatorial Mathematics.*

Thomas L. Saaty and Paul C. Kainen, *The Four-Color Problem.*

G. Pólya et al., *Notes on Introductory Combinatorics*, Modern Birkhäuser Classics,
DOI 10.1007/978-0-8176-4953-1_16, © Birkhäuser Boston, a part of Springer
Science+Business Media, LLC 2010

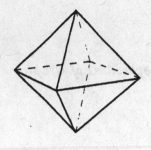

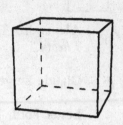

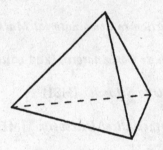

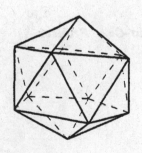

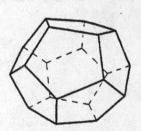